电力安全技术

——安全措施篇（中英双语）

主　编　李亚平　程　铭
副主编　刘晓菊　魏　娜　廖翔志
　　　　刘玲玲　张　斌

西南交通大学出版社
·成　都·

图书在版编目（CIP）数据

电力安全技术. 安全措施篇：汉文、英文 / 李亚平，程铭主编. -- 成都：西南交通大学出版社，2023.12
ISBN 978-7-5643-9594-0

Ⅰ. ①电… Ⅱ. ①李… ②程… Ⅲ. ①电力安全－教材－汉、英 Ⅳ. ①TM7

中国国家版本馆 CIP 数据核字（2023）第 231416 号

Dianli Anquan Jishu — Anquan Cuoshi Pian（Zhong-Ying Shuangyu）

电力安全技术——安全措施篇（中英双语）

主 编 / 李亚平 程 铭

责任编辑 / 张文越
封面设计 / GT 工作室

西南交通大学出版社出版发行
（四川省成都市金牛区二环路北一段 111 号西南交通大学创新大厦 21 楼 610031）
营销部电话：028-87600564 028-87600533
网址：http://www.xnjdcbs.com
印刷：四川玖艺呈现印刷有限公司

成品尺寸 185 mm × 260 mm
印张 16.25 字数 460 千
版次 2023 年 12 月第 1 版 印次 2023 年 12 月第 1 次

书号 ISBN 978-7-5643-9594-0
定价 98.00 元

课件咨询电话：028-81435775
图书如有印装质量问题 本社负责退换

前言

ABC

电能具有可大规模生产、传输、分配，转换方便，价格低廉等特点，因此在人类的日常生活和生产活动中得到广泛应用。然而各种安全隐患和事故也如影随形，威胁着人类的健康和生命。以电能的形式存在的电气安全危险因素便是其中之一。

电能看不见、听不到、嗅不着，对人类来说，其不具备直观识别的特性，不易识别；电以接近光的速度传播，事故的到来猝不及防；电气安全隐患和事故概念抽象，对非专业人员而言深奥难明。正因如此，对电气安全隐患和事故的防范机理和对策的研究非常重要，特别是在当今科学技术飞速发展的时代，电气设备以极快的速度更新换代，这就要求人类不但要有效防范各种已知的电气安全隐患和事故，还要及时学会应对新技术、新工艺、新方法带来的新隐患、新事故。

如何消除电气安全隐患、防范电气安全事故早已成为一个具有极为普遍、广泛和重要意义的长久课题。这也衍生出安全工程领域的一个重要分支——电力安全工程（电气安全工程）。“电力安全工程”作为一门课程，适用于电力专业的大学生、从事电力相关行业的技术人员。

国家教育事业发展“十四五”规划要求突出职业教育类型特色，深入推进改革创新，优化结构布局，推行“1+X”证书制度，深化产教融合等。本书充分体现了“十四五”规划的职业教育目标，将电力安全工程理论知识与实训内容相结合，构成理实一体化、模块化、任务化教学。本书侧重于讲解电力安全基础知识以及相关实训内容，具体包含电力安全工器具的检查、使用和保管，电力测量仪表的检查和使用，安全组织措施的应用，安全技术措施的应用等内容。

本教材由高级工程师、高级技师李亚平和程铭担任主编，刘晓菊、魏娜、廖翔志、刘玲玲、张斌担任副主编。其中李亚平编写模块一，程铭、刘晓菊共同编写模块二，魏娜、廖翔志共同编写模块三，刘玲玲、张斌共同编写模块四，全书由李亚平统稿。编写过程中作者参考了大量相关书籍，谨向这些书籍作者致谢，在编写过程中得到编写人员单位和评审人员的大力支持和帮助，在此一并表示由衷的感谢。由于编写时间仓促，经验不足，水平有限，虽经反复修改，仍难免有不妥和不足之处，恳请读者批评指正。

目　录

ABC

模块一　电力安全工器具的检查、使用和保管 …… 001

Module Ⅰ　Inspection, use, and safekeeping of electric power safety tools and instruments …… 002

任务一　认识电力安全工器具 …… 003

Task Ⅰ　Learn electric power safety tools and instruments …… 004

任务二　基本绝缘安全工器具的检查与使用 …… 006

Task Ⅱ　Inspection and use of basic insulation safety tools and instruments …… 014

任务三　辅助绝缘安全工器具的检查与使用 …… 024

Task Ⅲ　Inspection and use of auxiliary insulation safety tools and instruments …… 027

任务四　一般防护安全工器具的检查与使用 …… 031

Task Ⅳ　Inspection and use of general protection safety tools and instruments …… 042

任务五　电力安全工器具的保管与存放 …… 055

Task Ⅴ　Safekeeping and storage of electric power safety tools and instruments …… 059

任务六　电力安全工器具检查、使用及保管实训 …… 064

Task Ⅵ　Practical training of inspection, use, and safekeeping of electric power safety tools and instruments …… 071

模块二　电力测量仪表的检查和使用 …… 079

Module Ⅱ　Inspection and use of electric measuring instrument …… 080

任务一　万用表的检查和使用 …… 081

Task Ⅰ　Inspection and use of multimeter …… 096

任务二　钳形电流表的检查和使用 …… 114

Task Ⅱ　Inspection and use of clip-on ammeter …… 118

任务三　绝缘电阻表的检查和使用 …… 123

Task Ⅲ　Inspection and use of insulation resistance meter …… 127

任务四　接地电阻表的检查和使用 …… 132

Task Ⅳ　Inspection and use of grounding resistance meter …… 136

任务五　电力测量仪表的检查与使用实训 …… 142

Task Ⅴ　Practical training of inspection and use of electric power measuring instrument …… 144

模块三　安全组织措施的应用 …… 147
Module Ⅲ　Application of safety and organizational measures …… 149
任务一　现场勘察制度及其应用 …… 151
Task Ⅰ　On-site survey system and its application …… 152
任务二　工作票制度及其应用 …… 154
Task Ⅱ　Work ticket system and its application …… 160
任务三　工作许可制度及其应用 …… 168
Task Ⅲ　Work permit system and its application …… 169
任务四　工作监护制度及其应用 …… 171
Task Ⅳ　Work supervision system and its application …… 172
任务五　工作间断、转移和终结制度及其应用 …… 174
Task Ⅴ　The system of work interruption, transfer and completion and its application …… 176
任务六　电气设备现场勘察实训 …… 178
Task Ⅵ　Training on site survey of electrical equipment …… 183
任务七　变电站（发电厂）第一种工作票的填写实训 …… 189
Task Ⅶ　Training on filling of the first type of work ticket for substation (power plant) …… 194
任务八　变电站第一种工作票执行流程演练实训 …… 200
Task Ⅷ　Drill and training on the implementation procedure for the first type of work ticket for substation …… 202
模块四　安全技术措施的应用 …… 205
Module Ⅳ　Application of technical measures for safety …… 206
任务一　停　电 …… 207
Task Ⅰ　Power failure …… 209
任务二　验　电 …… 212
Task Ⅱ　Verification of live parts …… 215
任务三　接　地 …… 219
Task Ⅲ　Grounding …… 223
任务四　悬挂标示牌和装设遮栏（围栏） …… 229
Task Ⅳ　Hanging sign boards and mounting barriers (fences) …… 232
任务五　高压验电器的检查与使用实训 …… 236
Task Ⅴ　Inspection and training in using HV electroscope …… 239
任务六　装拆接地线实训 …… 243
Task Ⅵ　Training in mounting and removing grounding wires …… 246
任务七　悬挂标示牌和装设遮栏实训 …… 250
Task Ⅶ　Training in hanging sign boards and mounting barriers …… 252
参考文献 …… 254

模块一 电力安全工器具的检查、使用和保管

在电力网络建设、运行和维护工作中，现场作业人员经常使用高压验电器、绝缘操作杆（棒）、绝缘隔板、绝缘罩、携带型短路接地线等电力安全工器具。这些必要的工器具不仅能保证工作任务的完成，更能起到保护作业人员人身安全的作用，如防止人身触电伤害、高处坠落伤人、高处坠物伤人、电弧灼伤等。为充分发挥电力安全工器具的作用，电力从业人员应对各种电力安全工器具的基本结构、性能充分了解，并熟练掌握其使用和保管方法。

学习目标：

（1）能正确说出电力安全工器具的分类。

（2）能正确进行常用电力安全工器具使用之前的检查。

（3）能正确使用常用的电力安全工器具。

（4）能正确说出常用电力安全工器具使用过程中的注意事项。

（5）能正确说出常用电力安全工器具的保管方法。

Module Ⅰ Inspection, use, and safekeeping of electric power safety tools and instruments

In the construction, operation and maintenance of power network, on-site operators often use electric power safety tools and instruments, such as HV electroscope, insulating bar (rod), insulating barrier, insulating cover, and portable short circuit grounding wire. These necessary tools and instruments not only ensure the completion of working tasks, but also play a vital role in protecting the personal safety of operators, such as preventing electric shock injury, personal injury from falls at heights, personal injury caused by falling objects from heights, as well as electric arc burns. For the optimal utilization of electric power safety tools and instruments, electrical professionals should have a comprehensive understanding of their basic structures and performances. Additionally, they must be proficient in using and storing these tools and instruments effectively.

Learning objectives:

(1) Can correctly state the classification of electric power safety tools and instruments.

(2) Can correctly carry out the inspection before the use of commonly used electric power safety tools and instruments.

(3) Can correctly use commonly used electric power safety tools and instruments.

(4) Can correctly state the precautions during the use of commonly used electric power safety tools and instruments.

(5) Can correctly state the safekeeping methods for commonly used electric power safety tools and instruments.

任务一　认识电力安全工器具

一、电力安全工器具的作用

在电力行业根据专业和工种的不同，现场作业人员要为完成不同的工作任务而实施不同的操作。生产实践经验告诉我们，为了安全、保质、高效地完成工作任务，现场作业人员必须携带并正确使用各种电力安全工器具。例如，对运行中的电气设备进行巡视、改变运行方式、检修试验时，需要使用电力安全工器具；在线路施工中，离不开登高安全工器具；在带电的电气设备上或邻近带电设备的地方工作时，为了防止工作人员触电或被电弧灼伤，需使用绝缘安全工器具。所以，电力安全工器具是指用于防止触电、坠落、电弧灼伤等工伤事故，保障现场作业人员人身安全的各种专用工具，这些工具是现场作业人员日常作业中必不可少的。

二、电力安全工器具的分类

电力安全工器具分为绝缘安全工器具和一般防护安全工器具两大类。

（一）绝缘安全工器具

绝缘安全工器具又分为基本绝缘安全工器具和辅助绝缘安全工器具两类。

1. 基本绝缘安全工器具

基本绝缘安全工器具是指能直接操作带电设备、接触或可能接触带电体的工器具，如高压验电器、绝缘操作杆(棒)、绝缘隔板、绝缘罩、携带型短路接地线等。

2. 辅助绝缘安全工器具

辅助绝缘安全工器具是指绝缘强度不能承受设备或线路的工作电压，只是用于加强基本绝缘安全工器具的保护作用，用以防止接触电压、跨步电压、泄漏电流电弧等对操作人员的伤害，不能用辅助绝缘安全工器具直接接触高压设备的带电部分。常见的辅助绝缘安全工器具如绝缘手套、绝缘靴、绝缘胶垫、绝缘台等。

（二）一般防护安全工器具

一般防护安全工器具是指本身没有绝缘性能，但可以起到防护工作人员发生事故的用具。这种安全工器具主要用作防止检修设备误送电，防止工作人员走错间隔、误登带电设备，保证人与带电体之间的安全距离，防止电弧灼伤、高空坠落等事故。一般防护安全工器具包括安全帽、安全带、梯子、安全绳、脚扣、防静电服（静电感应防护服）、防电弧服、导电鞋（防静电鞋）、防护眼镜、过滤式防毒面具、安全围栏（网）和标示牌等。

Task Ⅰ Learn electric power safety tools and instruments

Ⅰ. Functions of electric power safety tools and instruments

In the power industry, according to different disciplines and types of work, on-site operators need to carry out different operations to complete different work tasks. Production practice tells us that in order to complete the task safely and efficiently with quality guaranteed, on-site operators must carry and correctly use various electric power safety tools and instruments. For example, when conducting routine inspections, changing operation modes, and conducting maintenance tests on electrical equipment in operation, it is necessary to use electric power safety tools and instruments. Safety tools and instruments for climbing are indispensable in the construction of lines. For working on charged electrical equipment or near charged equipment, insulation safety tools and instruments shall be used in order to prevent operators from electric shock or burned by arc. Therefore, electric power safety tools and instruments refer to a variety of special tools and appliances that are used to prevent industrial accidents such as electric shock, fall, and arc burn, and are used to ensure the personal safety of on-site operators. These tools and instruments are indispensable in the daily operation of on-site operators.

Ⅱ. Classification of electric power safety tools and instruments

Electric power safety tools and instruments are divided into two categories: insulation safety tools and instruments and general protection safety tools and instruments.

(Ⅰ) Insulation safety tools and instruments

Insulation safety tools and instruments can be divided into two categories: basic insulation safety tools and instruments and auxiliary insulation safety tools and instruments.

1. Basic insulation safety tools and instruments

Basic insulation safety tools and instruments refer to tools and instruments that can be used to directly operate charged equipment, to contact or may contact with charged bodies, such as HV electroscope, insulating bar (rod), insulating barrier, insulating cover, and portable short circuit grounding wire.

2. Auxiliary insulation safety tools and instruments

Auxiliary insulation safety tools and instruments refer to tools and instruments with insulating strength that cannot withstand the operating voltage of equipment or lines, but are only used to strengthen the security functions of basic insulation safety tools and instruments to prevent damage to operators caused by contact voltage, step voltage, leakage current arc, etc. Do not directly contact the live parts of HV equipment with auxiliary insulation safety tools and instruments. Common auxiliary insulation safety tools and instruments include insulating gloves, insulating boots, insulating rubber pad, and insulating stand.

(Ⅱ) General protection safety tools and instruments

General protection safety tools and instruments refer to tools and instruments that do not have insulation performance but can protect operators from accidents. Such safety tools and instruments are mainly used to prevent power transmission by mistake from equipment under maintenance, to prevent operators from entering wrong electrified compartment and boarding live equipment by mistake, to ensure safe distance between the person and the charged body, and to prevent accidents such as arc burns and high-altitude falling. General protection safety tools and instruments include safety helmets, safety belts, ladders, safety ropes, grapplers, anti-static clothing (static induction protective suits), arc preventive clothing, conductive shoes (anti-static shoes), goggles, filter gas masks, safety fences (nets) and sign boards.

任务二　基本绝缘安全工器具的检查与使用

本任务主要介绍绝缘操作杆（棒）、验电器、携带型短路接地线及个人保安接地线这 4 种基本绝缘安全工器具。通过学习，学习者应掌握其检查及使用方法。

一、绝缘操作杆（棒）

（一）绝缘操作杆（棒）的作用

绝缘操作杆（棒）简称为绝缘杆，是用于短时间对带电设备进行操作的绝缘工具，如接通或断开高压隔离开关、跌落式熔断器及拆除临时接地线等。

（二）绝缘操作杆（棒）的结构

绝缘操作杆（棒）由工作部分、绝缘部分、护环和握手部分组成，如图 1-1 所示。

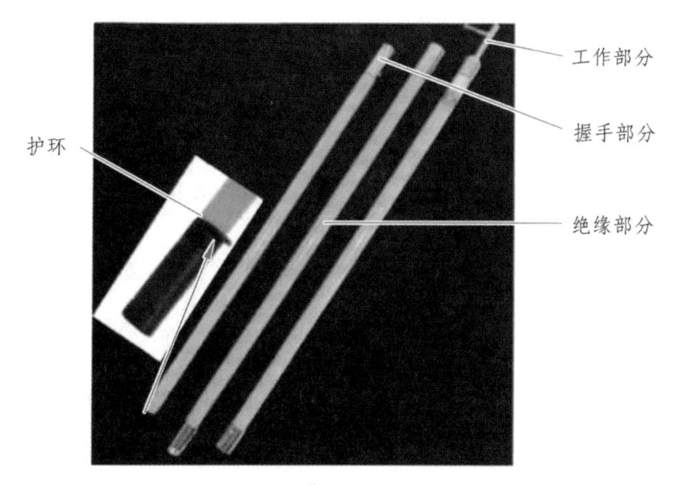

图 1-1　绝缘操作杆（棒）的结构

（1）工作部分：一般由金属或具有较大机械强度的绝缘材料（如玻璃钢）制成，一般不宜过长，在满足工作需要的情况下，长度不宜超过 8 cm，以免操作时发生相间或接地短路。

（2）绝缘部分：一般由合成材料制成，用于绝缘隔离。

（3）握手部分：一般由环氧树脂管制成，绝缘部分和握手部分两者之间由护环隔开。

（三）绝缘操作杆（棒）的检查

绝缘杆的检查事项如图 1-2 所示。

（1）电压等级。检查电压等级应与电气设备或线路的电压等级相符（等于或高于电气设备或线路的电压等级）。

（2）标签、试验合格证。检查标签、试验合格证是否齐全、完好，本次使用日期应在试验有效期内，否则不得使用。每年应进行一次预防性试验。

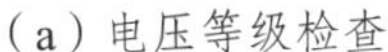

（a）电压等级检查　　（b）外观检查

图 1-2　绝缘杆的检查

（3）外观。检查并确保绝缘杆的外观清洁、干燥，杆身光洁、无裂纹或損伤，工作触头无破损，否则应禁止使用，绝缘部分和握手部分之间应有护环隔开，护环完好。

（四）绝缘操作杆（棒）的使用

（1）戴绝缘手套。使用时，应正确佩戴绝缘手套、穿绝缘靴（鞋），手握在绝缘杆的握手部分，不能超过护环。

（2）合适的站立位置。操作人应选好合适的站立位置，与相邻带电体保持足够的安全距离，避免物件失控落下时，造成人员伤亡；同时应注意防止绝缘杆被设备短接，保持有效的绝缘长度。

（3）雨天加装防雨罩。雨天在户外操作电气设备时，绝缘杆的绝缘部分应有防雨罩。防雨罩的上口应与绝缘部分紧密结合，确保无渗漏现象，以使罩下部分的绝缘保持干燥，如图 1-3 所示。

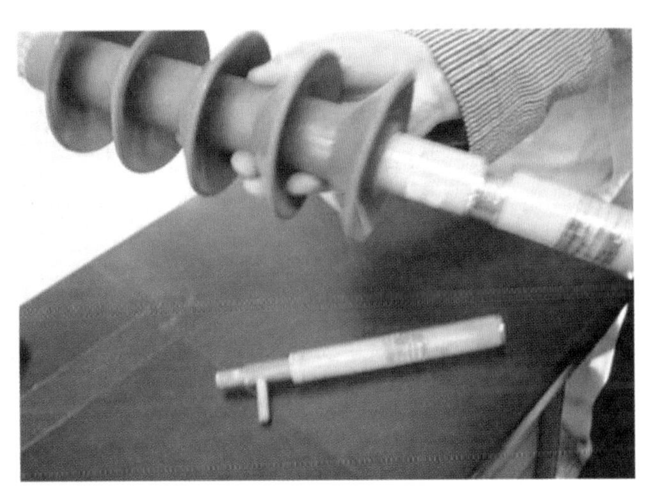

图 1-3　加装防雨罩的绝缘杆

二、验电器

（一）验电器的分类及工作原理

验电器也称携带型电压指示器，是一种用于检测设备或导线是否带电的轻便仪器，分为低压验电器和高压验电器两种。

1. 低压验电器

低压验电器又称为验电笔，是一种用氖灯指示是否带电的基本电气安全用具。为便于携带，它多被制成类似钢笔或螺丝刀的形状，如图 1-4 所示。

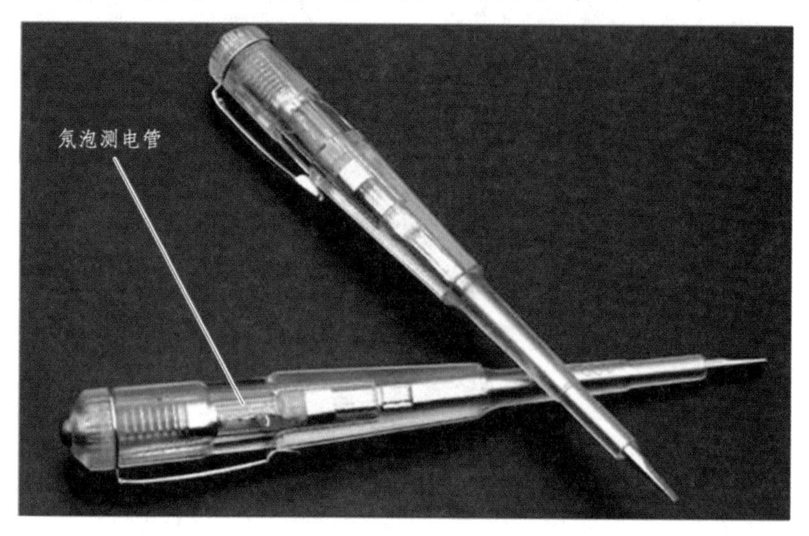

图 1-4 低压验电器（验电笔）

验电笔只能用于 380/220 V 的系统，使用前要在有电的设备或线路上试验一下，以检验其是否良好。使用时手拿验电笔，以一个手指触及验电笔后端的金属片，金属笔尖与被检查的带电部分接触，若设备带电，就有电流流过氖灯使氖灯发出亮光；如果氖灯不亮，说明不带电。灯越亮则电压越高，越暗则电压越低。

在三相四线制系统中，用验电笔触及中性线，若验电笔氖灯亮，说明系统故障或负荷不平衡。这是因为正常运行时，中性线上没有电压，但若发生相间短路、单相接地、相线断线或三相负荷不平衡，中性线上均会出现电压。

在三相三线制中，用验电笔分别触及三相时，验电笔氖灯在其中两相较亮，一相较暗，表明灯光暗的一相有接地现象。

2. 高压验电器

高压验电器又称测电器、试电器或电压指示器，是检验电气设备、电器、导线上是否有电的一种专用电力安全工器具。当每次断开电源进行检修时，必须先用它验明设备确实无电后，方可进行工作。

测电器接触电极接触到被试部位后，被测试部分的电信号传送到测试电路，被测试部分有电时测电器发出音响和灯光闪烁信号报警，无电时没有任何信号指示。为检查验电器工作是否正常，设有一试验开关，按下后若发出声响和灯光信号，则表示测电器工作正常。

（二）高压验电器的结构

高压验电器由握手部分、抽拉式绝缘杆、蜂鸣器、接触电极、护环等部分组成，其结构如图 1-5 所示。

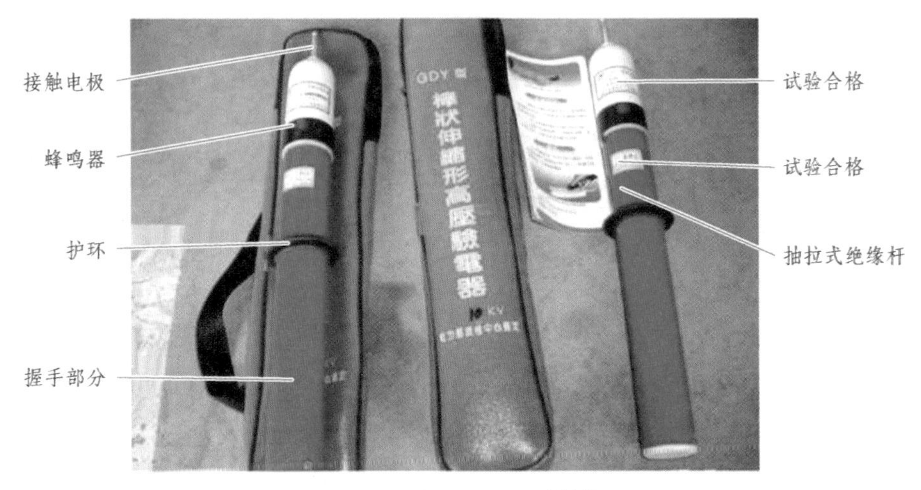

图 1-5　高压验电器的结构

（三）高压验电器的检查

高压验电器的检查事项如下：

（1）电压等级。使用前，按被测设备的电压等级，选择同等电压等级的高压验电器，禁止使用电压等级不对应的高压验电器进行验电，以免现场测验得出错误的判断，如图 1-6 所示。

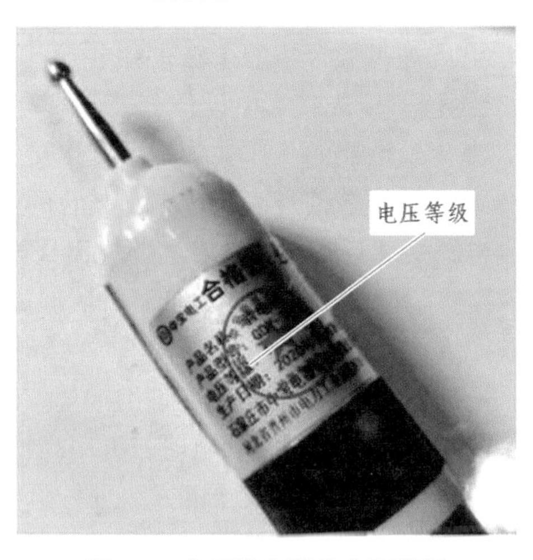

图 1-6　高压验电器的电压等级

（2）标签、试验合格证。检查并确保高压验电器标签完好，标签上应写明高压验电器的编号（110 kV 及以上电压等级高压验电器标签上还需写明配用的绝缘杆节数）。检查工作触头、绝缘杆的试验合格证是否齐全，如图 1-7 所示。本次使用日期应在试验有效期内，否则不得使用，其试验周期为 1 年。

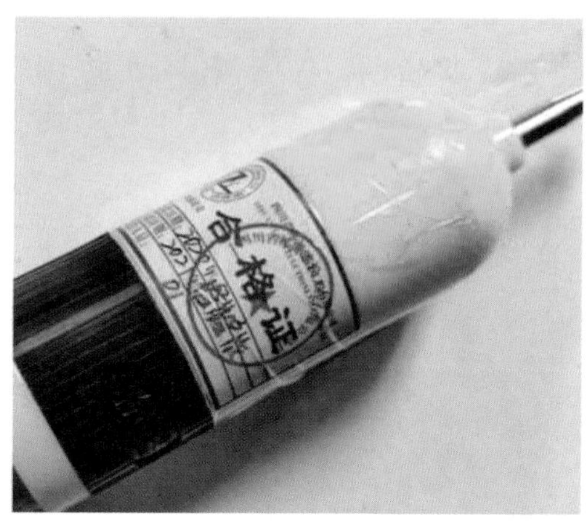

图 1-7　验电器的试验合格证

（3）性能、外观。使用前，应检查高压验电器外观并确保其光滑、干燥、无破损、有护环。使用抽拉式高压验电器时，绝缘杆应完全拉开检查，各节应配合合理，拉伸后不应自动回缩。按压工作触头，初步检查高压验电器声光信号应正常。

（4）接触电极。接触电极应无烧焦、熔化、锈蚀、断裂的痕迹。

（四）高压验电器的使用

（1）使用高压验电器时，操作人应戴绝缘手套、穿绝缘靴（鞋），手握在护环下侧握手部分且不得超过护环。

（2）验电时，应使用相应电压等级且合格的接触式验电器，在装设接地线或合接地刀闸（装置）处对各相分别验电。验电前，应先在有电设备上进行试验，并确认验电器良好。

（3）在有电设备验电时，高压验电器的工作触头不能直接接触带电体，只能逐渐接近带电体，直至验电器发出声、光或其他报警信号为止。对于被验的无电设备，验电器的工作触头应直接接触。应注意防止验电器受邻近带电体的影响而发出报警信号。

（4）同杆架设多层电力线路进行验电时，应先验低压，后验高压；先验下层，后验上层；先验距人体较近的导线，后验距人体较远的导线。

（5）高压验电器不得在雷、雨、雪等恶劣天气时使用。

（6）验电前，验电器无法在有电设备上进行验电时，可用高压发生器等确认验电器良好。

（7）330 kV 及以上的电气设备，可采用间接验电方法进行验电。

三、携带型短路接地线

（一）携带型短路接地线的作用

（1）当对高压设备进行停电检修或进行其他工作时，防止检修设备突然来电。

（2）防止邻近带电高压设备产生的感应电压对工作人员造成伤害。

（3）放尽剩余电荷。

（二）携带型短路接地线的结构

携带型短路接地线通常由线夹、操作棒、握手部分、汇流夹、多股软铜线和接地端线夹等组成，其结构如图 1-8 所示。

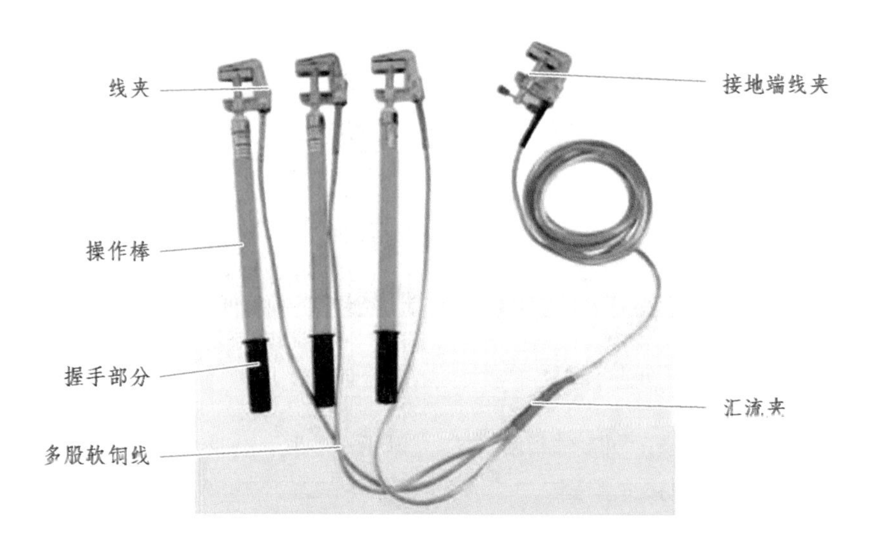

图 1-8　携带型接地线的结构

（1）线夹：起到接地线与设备的可靠连接作用。

（2）多股软铜线：应承受工作地点流过的最大短路电流，同时应有一定的机械强度，截面积不得小于 25 mm^2。多股软铜线的透明塑料外套起保护作用。

（3）接地端线夹：起到接地线与接地网的连接作用，一般是用螺钉紧固，或用接地棒打入地下，打入地下时深度不得小于 0.6 m。

（4）汇流夹：汇集短接三相短路电流。

（三）携带型接地线的检查

携带型接地线的检查事项如下：

1. 电压等级

使用前，应检查接地线的电压等级必须符合接地设备电压等级，切不可任意取用。

2. 标签、试验合格证

检查接地线有标签、试验合格证。本次使用日期应在试验有效期内，否则不能使用。其成组直流电阻的试验周期不超过 5 年，其操作棒的试验周期是 5 年。其试验合格证及标签如图 1-9 所示。

3. 连接部件及外观

（1）各连接处接触良好，螺栓紧固，无松动、滑丝、锈蚀、融化现象。

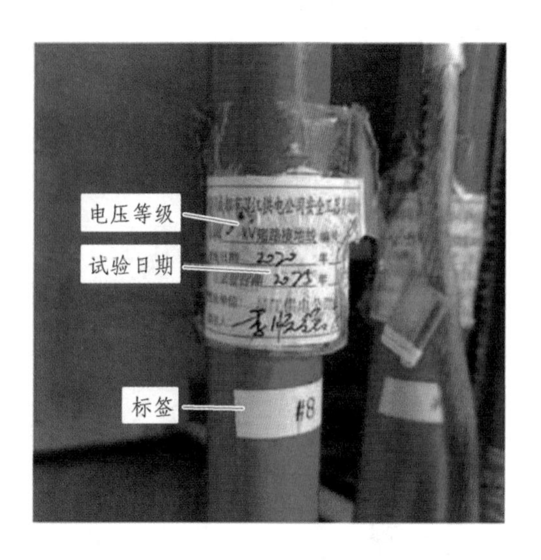

图 1-9　接地线的试验合格证

（2）多股软铜线护套应无严重磨损，铜线应无断股、散股、松股，其截面积不小于 25 mm²，接地铜线和三根短铜线的连接应牢固。

（3）夹具完好无裂纹，夹具或线钩弹力正常。

（4）绝缘操作棒表面清洁光滑，无气泡、皱纹、开裂、划伤，绝缘漆无脱落。

（5）握手部分与绝缘杆、绝缘杆与金属件的连接应牢固可靠，有护环且护环完好。

（四）携带型接地线的使用

（1）装接地线之前必须验电，验电位置必须与装设接地线的位置相符。

（2）接地时要两人进行，一人操作，一人监护。

（3）操作人应选好合适的站立位置，保证与相邻带电体足够的安全距离。

（4）为防止接地时绝缘操作棒受潮产生的泄漏电流危及操作人员的安全，在使用时必须戴绝缘手套、穿绝缘靴（鞋），手握在绝缘操作棒的护环以下部位。

（5）装、拆接地线时，人体不得碰触接地线，并按正确的顺序操作：要先接接地端，后接导线端；先挂低压，后挂高压；先挂下层，后挂上层。拆接地线时的顺序与此相反。

若杆塔无接地引下线时，可采用临时接地棒，接地棒打入地下的深度不得小于 0.6 m。如利用铁塔接地时，可每相分别接地，但铁塔与接地线连接部分应清除油漆。

接地完毕后，必须检查接地线的线夹与导体接触良好，并有足够的夹紧力，以防通过短路电流时，由于接触不良而熔断或因电动力的作用而脱落。

四、个人保安接地线

个人保安接地线如图 1-10 所示，是为防止邻近带电设备和线路产生的感应电压，保证操作人员的人身安全而装设的接地装置。个人保安接地线仅作为预防感应电使用，不得以此代替工作接地线。只有在工作接地线挂好后，方可在工作相上挂个人保安接地线。

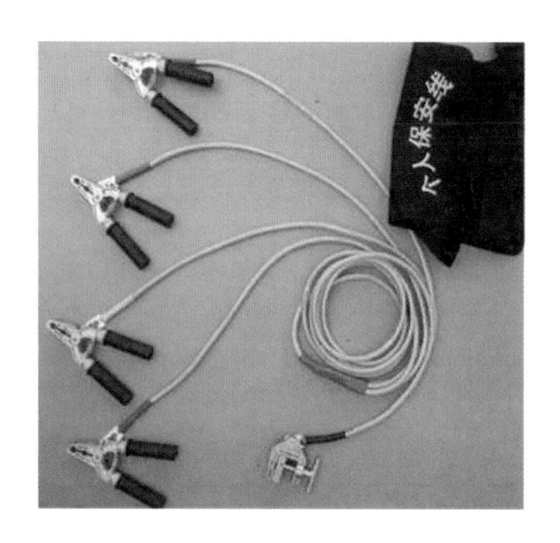

图 1-10　个人保安接地线

个人保安接地线的成组直流电阻试验周期不超过 5 年。

个人保安接地线由工作人员自行携带，自装自拆。凡在 110 kV 及以上同杆塔并架或相邻的平行有感应电的线路上停电工作时，应在工作相上使用个人保安接地线，并不准采用搭连虚接的方法接地。工作结束时，工作人员应拆除所挂的个人保安接地线。

Task Ⅱ Inspection and use of basic insulation safety tools and instruments

In this task, 4 types of basic insulation safety tools and instruments, namely, insulating bar (rod), electroscope, portable short circuit grounding wire, and personal safety grounding wire, are introduced. Master their inspection and using methods through study.

Ⅰ. Insulating bar (rod)

(Ⅰ) Functions of insulating bar (rod)

Insulating bar (rod), referred to as insulating rod, is an insulation tool used to operate live equipment for a short time, such as turning on or off HV isolating switch, drop-out fuse, and removing temporary grounding wire.

(Ⅱ) Structure of insulating bar (rod)

The insulating bar (rod) is composed of working part, insulating part, protective ring, and handle part, as shown in Fig. 1-1.

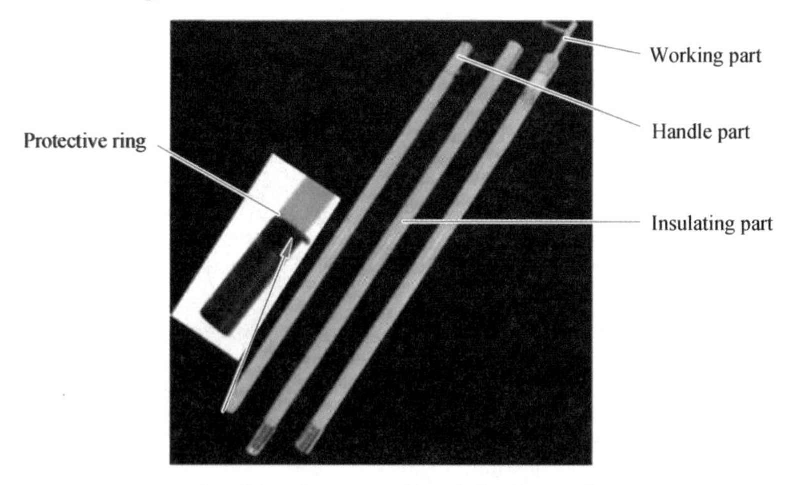

Fig. 1-1 Structure of insulating bar (rod)

(1) Working part: It is generally made of metal or insulating materials with high mechanical strength (such as fiberglass), it should not be too long. In order to meet the work needs, the length should not exceed 8 cm to avoid interphase or grounding short circuit during operation.

(2) Insulating part: It is generally made of synthetic materials, used for insulation and isolation.

(3) Insulating part and handle part: They are generally made of epoxy resin tubes, and the two are separated by a protective ring.

(Ⅲ) Inspection of insulating bar (rod)

The inspection items of the insulating rod are as shown in Fig. 1-2.

(1) Voltage class. Check that the voltage class shall be in line with the voltage class of the electrical equipment or line (equal to or higher than the voltage class of the electrical equipment or line).

(2) Label and test certificate. Check whether the label and test certificate are complete and in good condition, and the date of use shall be within the validity period of the test, otherwise it shall not be used. Preventive testing shall be conducted once a year.

(a) Voltage class inspection

(b) Visual inspection

Fig. 1-2 Inspection of insulating rod

(3) Appearance. Check whether the insulating rod is clean, dry, smooth, crack-free, or undamage. The working contact terminal shall not be damaged, otherwise it shall be prohibited to use. There is a protective ring between the insulating part and the handle part, and the protective ring is intact.

(Ⅳ) Use of insulating bar (rod)

(1) Wear insulating gloves. The user shall wear insulating gloves and insulating boots (shoes) correctly, and the hand shall hold the handle part of the insulating rod, not exceeding the protective ring.

(2) Suitable standing position. The operator shall choose a suitable standing position and maintain a sufficient safe distance from the adjacent charged body, so as to avoid casualties when the object is out of control and falls. At the same time, attention shall be paid to prevent the insulating rod from being short connected by the equipment and effective insulation length should be kept.

(3) Install a rain cover in rainy days. For operating electrical equipment outdoors on rainy days, the insulating part of the insulating rod shall have a rain cover. The upper opening of the rain cover

shall be tightly integrated with the insulating part, leakage is not allowed, in order to keep the insulation under the cover dry, as shown in Fig. 1-3.

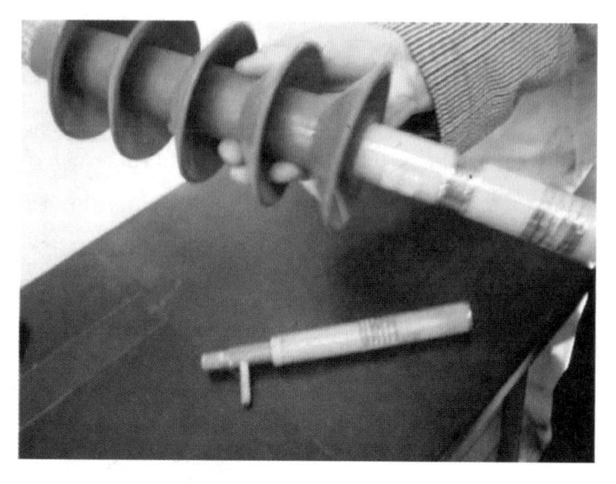

Fig. 1-3 Insulating rod with a rain cover

Ⅱ. Electroscope

(Ⅰ) Classification and working principle of electroscope

The electroscope, also known as the portable voltage indicator, is a portable instrument used to detect whether the equipment or conductor is charged. Electroscopes are divided into two types: LV electroscope and HV electroscope.

1. LV electroscope

LV electroscope, also known as electroprobe, is a basic electric safety appliance that uses a nitrogen lamp to indicate whether it is charged. For ease of carrying, it is often made into a shape similar to a pen or screwdriver, as shown in Fig. 1-4.

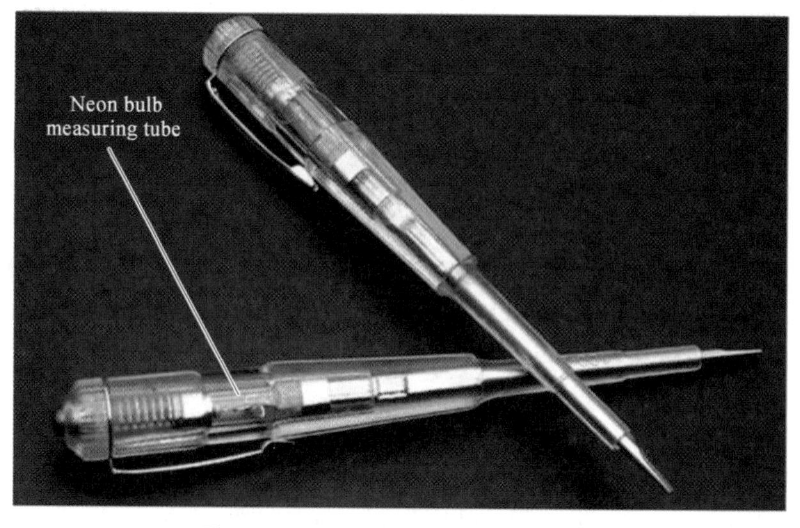

Fig. 1-4 LV electroscope (electroprobe)

The electroprobe can only be used in a 380/220 V system. It shall be tested on a live equipment or circuit before use to see if it is in good condition. When using, hold the electroprobe in hand and touch the metal strip at the back of the electroprobe with one finger. The metal tip of the electroprobe comes into contact with the charged part being inspected. If the equipment is charged, there is current flowing through the neon lamp, which emits a bright light; If the neon lamp is not on, it means the equipment is not charged. The brighter the lamp, the higher the voltage, and the darker the lamp, the lower the voltage.

In a three-phase four-wire system, touch the neutral wire with an electroprobe. If the neon lamp on the electroprobe is on, it indicates a system fault or load imbalance. This is because, during normal operation, there is no voltage on the neutral wire, but if there is an interphase short circuit, single-phase grounding, phase line break, or three-phase load imbalance, there will be voltage on the neutral wire.

In the three-phase three-wire system, when touching the three phases separately with an electroprobe, the neon lamp of the electroprobe is brighter in two phases and darker in one phase, indicating that the phase with dim light is grounded.

2. HV electroscope

HV electroscope, also known as electricity measurer, electricity tester, or voltage indicator, is a special electric power safety tool to check whether there is electricity on electrical equipment, appliance, and conductor. When each time the power supply is disconnected for maintenance, it must be used to verify that the equipment is indeed without power before it can work

After the contact electrode of the electricity measurer comes into contact with the tested part, the electric signal of the tested part is transmitted to the test circuit. When the tested part is charged, the tester emits sound and light flashing signals to alarm, but there is no signal indication when the tested part is not charged. In order to check whether the electroscope is working properly, there is a test switch. If the sound and light signals are sent out after pressing the switch, it means that the electroscope is working normally.

(Ⅱ) Structure of HV electroscope

HV electroscope consists of handle part, a pull-out insulating rod, a buzzer, a contact electrode, a protective ring and other parts, and its structure is as shown in Fig. 1-5.

(Ⅲ) Inspection of HV electroscope

The inspection items of the HV electroscope are as follows:

(1) Voltage class. Before use, select a HV electroscope with the same voltage class according to the voltage class of the equipment tested. It is forbidden to use a HV electroscope with different voltage class for electricity detection, so as not to make a wrong judgment in the field test, as shown in Fig. 1-6.

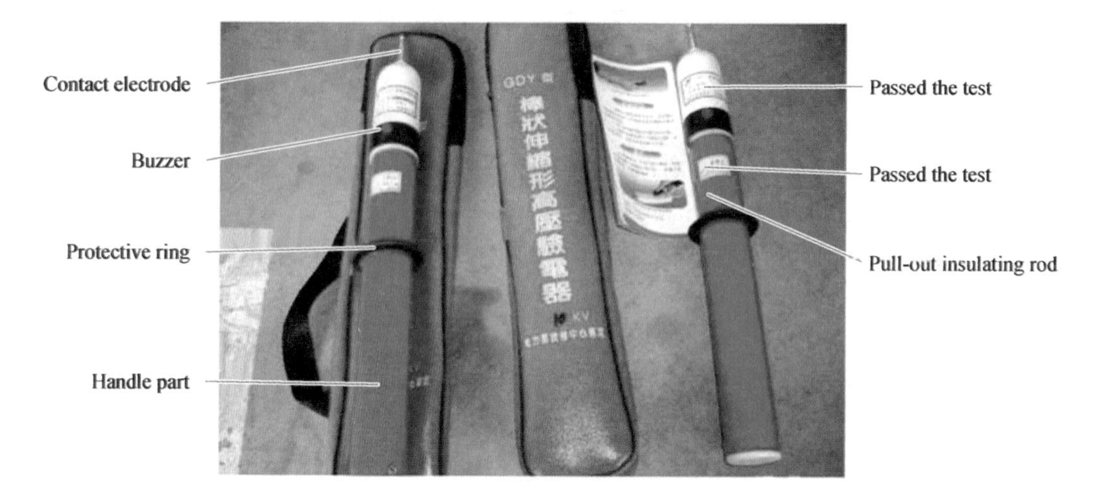

Fig. 1-5 Structure of HV electroscope

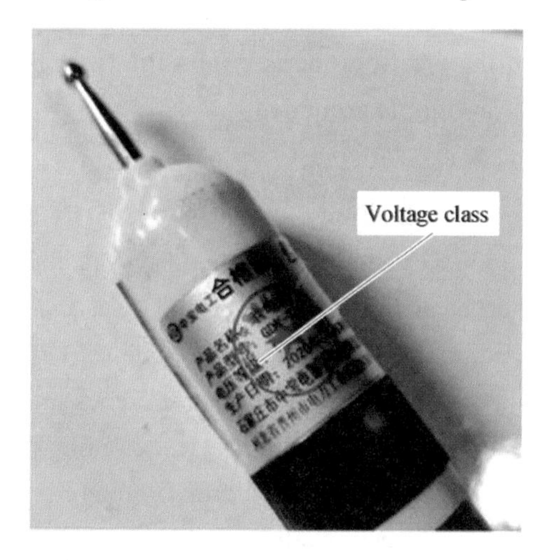

Fig. 1-6 Voltage class of HV electroscope

(2) Label and test certificate. Check whether the label of HV electroscope is intact. The number of the HV electroscope shall be indicated on the label (the number of insulating rods to be used shall also be indicated on the label of the insulating rod with a voltage class of 110 kV and above). Check whether the test certificates of working contact terminal and insulating rod are complete, as shown in Fig 1-7. The date of use shall be within the period of validity of the test, otherwise it shall not be used, and the test period is 1 year.

(3) Performance and appearance. Before use, check whether HV electroscope is smooth, dry, free from damage, and with a protective ring. When using the pull-out HV electroscope, the insulating rod shall be fully pulled open for inspection, each section shall fit reasonably, and shall not retract automatically after tension. Press the working contact terminal and initially check that the acoustic-optical signal of the HV electroscope shall be normal.

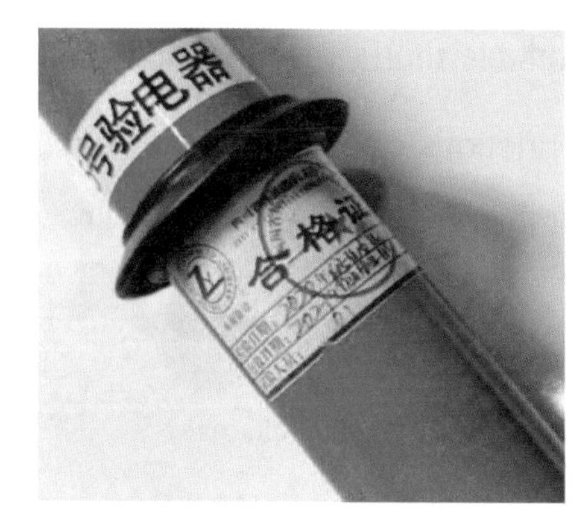

Fig. 1-7 Test certificate of electroscope

(4) Contact electrode. The contact electrode shall have no signs of scorch, melting, corrosion, or fracture.

(Ⅳ) Use of HV electroscope

(1) When using a HV electroscope, the operator shall wear insulating gloves and insulating boots (shoes), hold the handle part on the lower side of the protective ring, not exceeding the protective ring.

(2) During the electricity detection, the qualified contact electroscope with corresponding voltage class shall be used to inspect the electricity of each phase separately at the place where the grounding wire is installed or the grounding knife-switch (device) is closed. Before the electricity detection, the test shall be carried out on the charged equipment to make sure that the electroscope is in good condition.

(3) During the electricity detection of the charged equipment, the working contact terminal of the HV electroscope cannot directly contact the charged body, but can only gradually approach the charged body until the electroscope emits sound, light, or other alarm signals. For the tested uncharged equipment, the working contact terminal of the electroscope shall be in direct contact. Attention shall be paid to prevent electroscope from sending out alarm signals due to the influence of adjacent charged bodies.

(4) For the electricity detection of multi-layer electric power lines erected on the same pole, LV shall be tested first and HV later. the lower layer shall be tested first and upper layer later. The conductor closer to the human body shall be tested first, and the conductor further away from the human body later.

(5) HV electroscope shall not be used in severe weather such as thunder, rain, and snow.

(6) Before the electricity detection, if the electroscope is unable to conduct the electricity detection on the charged equipment, the HV generator can be used to confirm that the electroscope is in good condition.

(7) Electrical equipment of 330 kV and above can be tested by the indirect electricity detection method.

Ⅲ. Portable short circuit grounding wire

(Ⅰ) Functions of portable short circuit grounding wire

(1) When conducting interruption maintenance or other work on HV equipment, prevent the equipment under maintenance suddenly charged.

(2) Prevent the induced voltage generated by the adjacent charged HV equipment from causing harm to the operators.

(3) Discharge all residual charges.

(Ⅱ) Structure of portable short circuit grounding wire

Portable short circuit grounding wire is usually composed of wire clamp, operating rod, handle part, current collecting clamp, multi-strand soft copper wire, and grounding terminal wire clamp, etc. Its structure is as shown in Fig. 1-8.

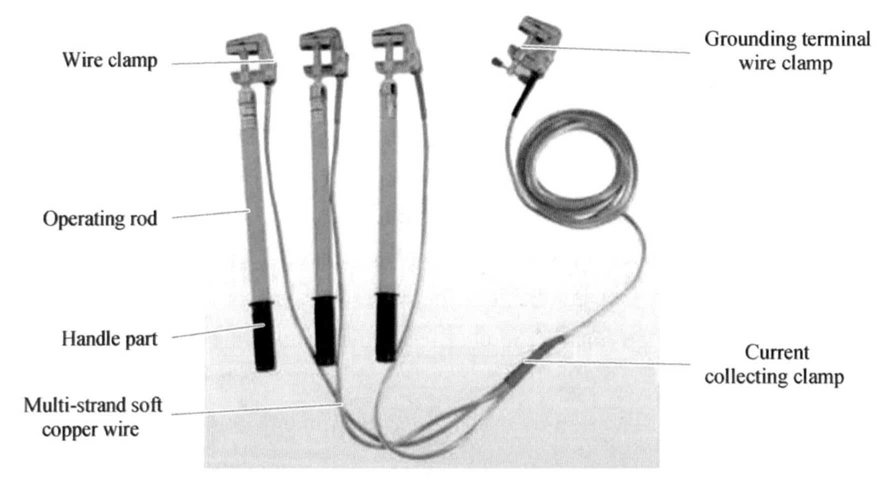

Fig. 1-8 Structure of portable grounding wire

(1) Wire clamp: It plays the role of reliable connection between grounding wire and equipment.

(2) Multi-strand soft copper wire: It shall withstand the maximum short circuit current flowing through the working site, and shall have a certain mechanical strength, with a cross sectional area not less than 25 mm^2. The transparent plastic jacket of multi-strand soft copper wire plays a protective role.

(3) Grounding terminal wire clamp: It serves as a connection between the grounding wire and the grounding grid, usually fastened with screws or driven underground with a grounding rod, with a depth of no less than 0.6 m.

(4) Current collecting clamp: Collect short connected three-phase short circuit current.

(Ⅲ) Inspection of portable grounding wire

The inspection items of the portable grounding wire are as follows:

1. Voltage class

Before use, the voltage class of the grounding wire must be checked to ensure that it meets the voltage class of the grounding equipment and cannot be arbitrarily used.

2. Label and test certificate

Check that the grounding wire has a label and test certificate. The date of use shall be within the period of validity of the test, otherwise it shall not be used. The test period of the group DC resistor is no more than 5 years, and the test period of the operating rod is 5 years. The test certificate and label are as shown in Fig. 1-9.

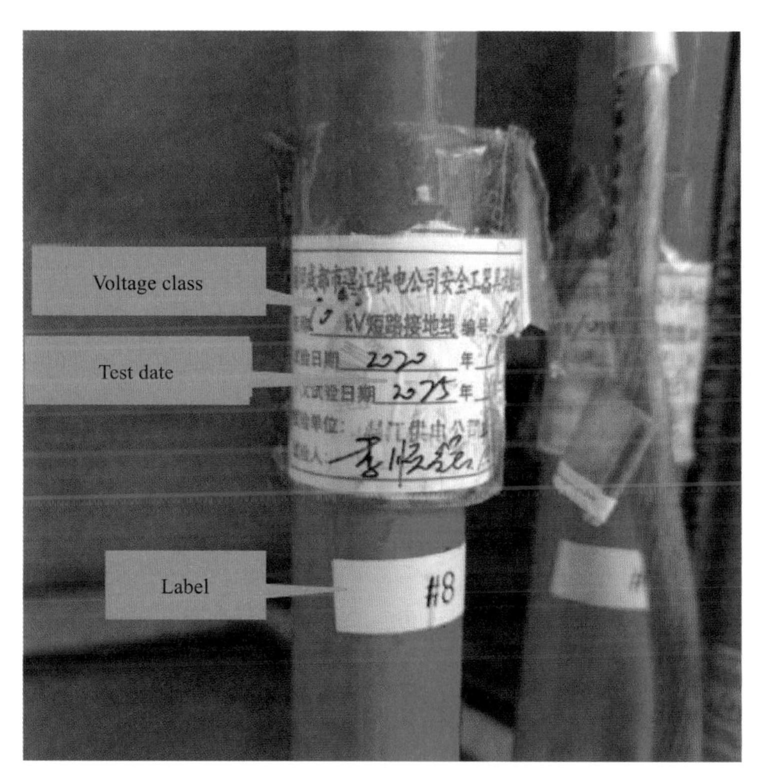

Fig. 1-9 Test certificate of grounding wire

3. Connecting component and appearance

(1) All connections have good contact, and the bolts are tightened, without looseness, slipping, rust, or melting.

(2) The sheath of multi-strand soft copper wire shall not be seriously worn. The copper wire shall have no broken, separate, or loose strands, and its cross sectional area shall not be less than 25 mm^2. The connection between the grounding copper wire and three short copper wires shall be firm.

(3) The grip shall be intact without cracks, and the elasticity of the grip or wire hook shall be normal.

(4) The surface of insulating rod shall be clean and smooth, with no bubbles, wrinkles, cracks, and scratches, and the insulation paint shall not fall off.

(5) The connection between the handle part and the insulating rod, the insulating rod and the metal part shall be firm and reliable, with protective rings which shall be intact.

(Ⅳ) Use of portable grounding wire

(1) Electricity detection must be done before the grounding wire is installed, and the position of the electricity detection must be consistent with the position of the grounding wire installed.

(2) Grounding requires two persons, one to operate and one to monitor.

(3) The operator shall choose a suitable standing position to ensure a sufficient safe distance from the adjacent charged body.

(4) To prevent leakage current caused by moisture on the insulating rod during grounding from endangering the safety of operators, it is necessary to wear insulating gloves, insulating boots (shoes), and hold the insulating rod below the protective ring.

(5) When installing or disassembling the grounding wire, the human body shall not touch the grounding wire and the operator shall operate in the correct order: the grounding terminal shall be connected first, and then the conductor terminal; first LV, then HV; first lower layer, then upper layer. The sequence to disassemble a grounding wire is the opposite.

If there is no grounding down conductor for the tower, a temporary grounding rod can be used, and the depth of the grounding rod into the ground shall not be less than 0.6 m. If the tower is grounded, each phase can be grounded separately, but the paint shall be removed from the connection between the tower and the grounding wire.

After the grounding is completed, it must be checked that the clamp of the grounding wire is in good contact with the conductor and has sufficient clamping force to prevent it from fusing due to poor contact or falling off due to the action of electric force when short circuit current passes through.

Ⅳ. Personal safety grounding wire

The personal safety grounding wire is as shown in Fig. 1-10. It is a grounding device installed to prevent induced voltage generated by adjacent charged equipment and lines, and ensure the personal safety of operators. The personal safety grounding wire is only used for the prevention of induced voltage and shall not be used as a substitute for the working grounding wire. Only if the work grounding wire is installed, the personal safety grounding wire be installed on the work phase.

The group DC resistance test period of personal safety grounding wire shall not exceed 5 years.

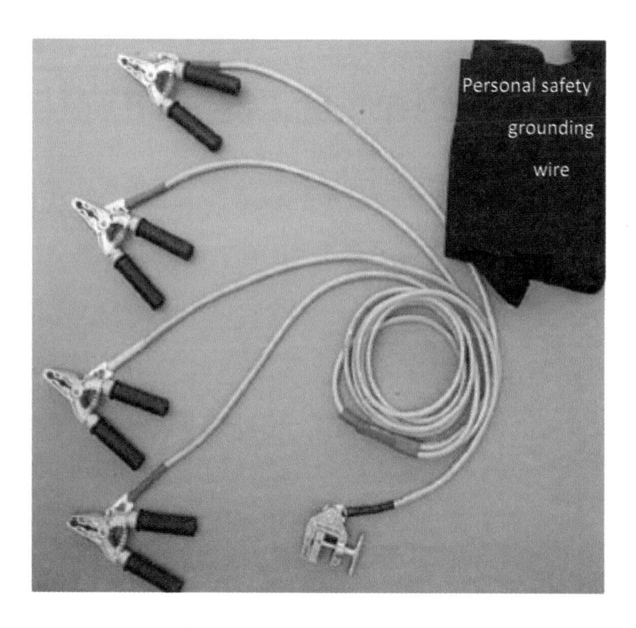

Fig. 1-10 Personal safety grounding wire

The personal safety grounding wire is carried installed and dismantled by the operators themselves. For interruption maintenance on 110 kV and above parallel or adjacent lines with induced electricity on the same tower, personal safety grounding wire shall be used on work phase, and it is not allowed to use the method of virtual connection and overlapping for grounding. At the end of the work, the operators shall remove the personal safety grounding wires that have been hung.

任务三　辅助绝缘安全工器具的检查与使用

本任务主要介绍绝缘手套、绝缘靴两种辅助绝缘安全工器具。通过学习，学习者应了解其用途、检查方法和使用方法。

一、绝缘手套

（一）绝缘手套的主要用途

绝缘手套是在高压设备上进行操作时使用的辅助安全工器具。例如操作高压隔离开关、装拆接地线、在高压回路上验电等需使用绝缘手套。绝缘手套也可以在低压下作为基本安全工器具使用。

（二）绝缘手套的检查

绝缘手套的检查项目如下：

（1）标签、试验合格证。使用前应检查标签、试验合格证是否齐全，如图 1-11（a）所示。标签上应写明本双绝缘手套的编号、左右手。本次使用日期应在试验有效期内，其试验周期为半年。

（2）外观。使用前应检查并确保其外观清洁、干燥，无发黏、裂纹、破口、气泡、发脆，手指有无粘连等损坏现象。

（3）内部。把手伸进去摸一下，内部应无潮湿现象。

（4）气密性。使用前应做充气试验。将绝缘手套朝手指方向卷曲进行充气试验，检查有无沙眼漏气现象，如图 1-11（d）所示，不能用图 1-11（b）、（c）所示卷曲方向进行气密性检查。

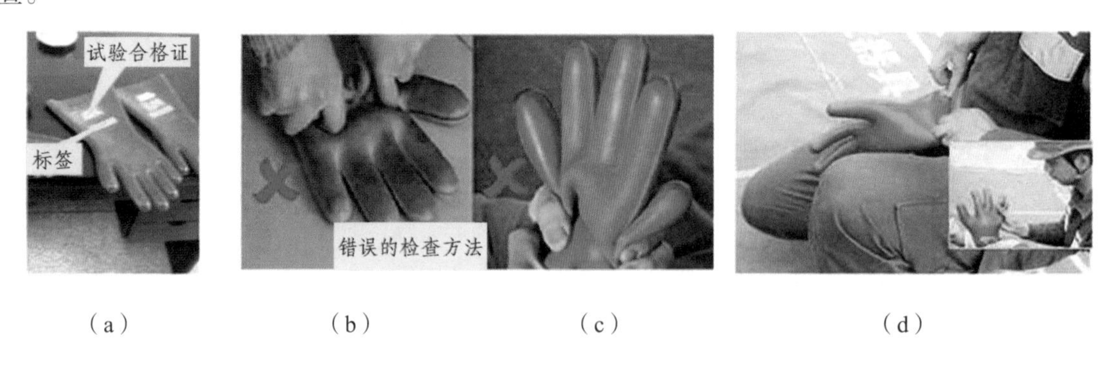

（a）　（b）　（c）　（d）

图 1-11　绝缘手套的检查

（三）绝缘手套的使用

（1）穿戴要求。戴绝缘手套时应将上衣袖口套入手套筒口内，如图 1-12 所示。

图 1-12　绝缘手套的使用

（2）使用场所。进行设备验电，装拆接地线，插上、取下合闸电源熔断器（电磁操动机构），用绝缘杆拉合隔离开关，经传动机构拉合断路器和隔离开关，高压设备发生接地时接触设备的外壳和构架，或其他有可能接触带电体时应戴绝缘手套。

（3）严禁把绝缘手套垫在操作手柄上使用。

（4）绝缘手套由特种橡胶制成，其长度应超过手腕 10 cm 以上。任何普通的、医疗用的、化学用的手套不能代替绝缘手套使用，也不得把绝缘手套做其他用途。

二、绝缘靴（鞋）

（一）绝缘靴（鞋）的作用

绝缘靴（鞋）的作用是使人体与地面绝缘，是在任何电压等级的地面上工作时，用来与地面保持绝缘和防止跨步电压触电（见图 1-13）的一种安全用具。

图 1-13　跨步电压触电

（二）绝缘靴（鞋）的检查

绝缘靴（图 1-14），其检查事项如下：

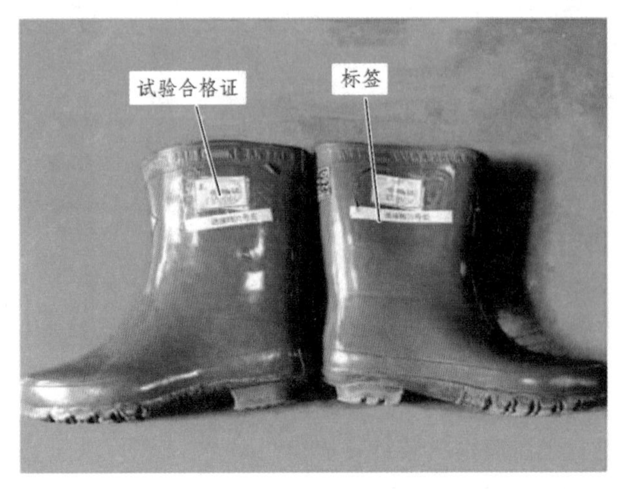

图 1-14　绝缘靴（鞋）

（1）使用前应检查标签、试验合格证是否齐全。标签上应写明本双绝缘靴（鞋）的编号、左右脚、鞋码，本次使用日期应在试验有效期内，其试验周期为半年。

（2）使用前应检查外观并确认是否干燥、清洁，是否无外伤、裂纹、破洞、毛刺、划痕、气泡、灼伤痕迹等。

（3）触摸绝缘靴，内部应无潮湿现象。

（4）检查绝缘靴大底应无磨损，无黄色面胶露出。

（三）绝缘靴（鞋）的使用

（1）使用绝缘靴时，应将裤管套入靴筒内，如图 1-15 所示。

（2）避免接触尖锐的物体，避免接触高温或腐蚀性物质，防止绝缘靴（鞋）受到损伤。

（3）雷雨天气或一次系统有接地时，巡视室内外高压设备应穿绝缘靴。接地网电阻不符合要求时，晴天也应穿绝缘靴。

（4）绝缘靴由特种橡胶制成，不能用胶鞋代替绝缘靴，严禁将绝缘靴挪作他用。

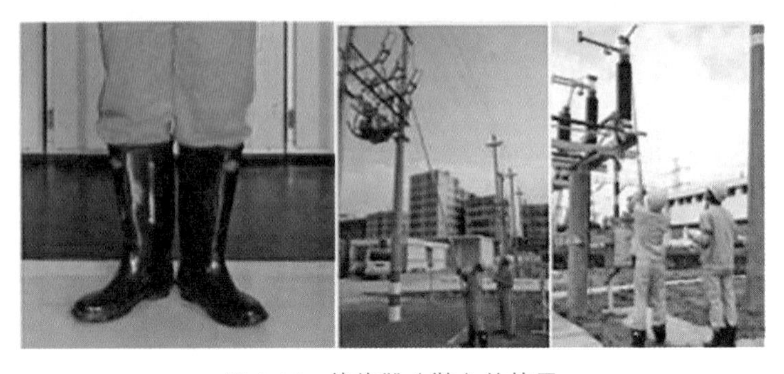

图 1-15　绝缘靴（鞋）的使用

Task Ⅲ　Inspection and use of auxiliary insulation safety tools and instruments

In this task, two types of auxiliary insulation safety tools and instruments, insulating gloves and insulating boots are mainly introduced. Learn their applications, inspection methods, and usage methods through study.

Ⅰ. Insulating gloves

(Ⅰ) Main applications of insulating gloves

Insulating gloves are auxiliary safety tools and instruments which are used when operating on HV equipment. For example, insulating gloves shall be used to operate HV isolating switch, install and disassemble grounding wire, conduct electricity detection on HV circuit and so on. Insulating gloves can also be used as basic safety tools and instruments under LV conditions.

(Ⅱ) Inspection of insulating gloves

The inspection items of insulating gloves are as in Fig. 1-11:

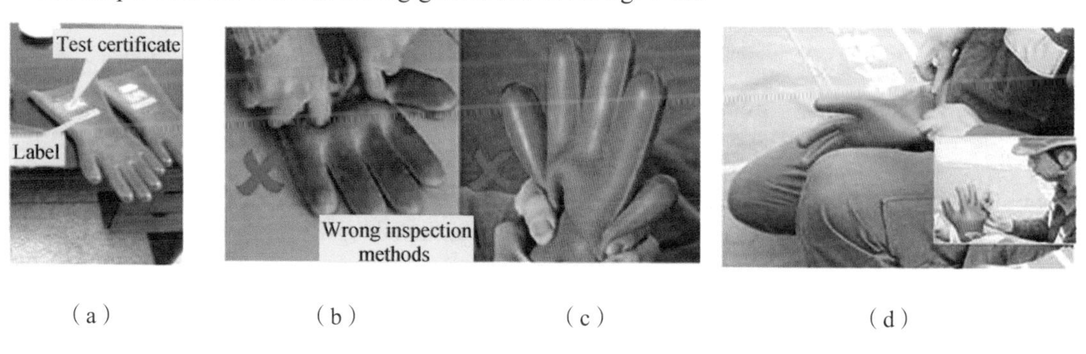

(a)　(b)　(c)　(d)

Fig. 1-11　Inspection of insulating gloves

(1) Label and test certificate. Check whether the label and test certificate are complete before use, as shown in Fig. 1-11(a). The label shall clearly indicate the number of this pair of insulating gloves, which one is for left hand and which one is for right hand. The date of use shall be within the period of validity of the test, and the test period is half a year.

(2) Appearance. Before use, the appearance shall be checked to guarantee cleanliness, dryness, no stickiness, cracks, breaks, bubbles, brittleness, and no damage such as finger adhesion.

(3) Interior. Put a hand in and touch it, and the interior shall not be damp.

(4) Gas tightness. Air inflation test shall be conducted before use. Inflate the insulating gloves by curling toward the fingers to check for small holes and air leakage, as shown in Fig. 1-11(d). The air tightness inspection cannot be carried out using the curling direction shown in Fig. 1-11 (b) and Fig. 1-11 (c).

(Ⅲ) Use of insulating gloves

(1) Wearing requirements. When wearing insulating gloves, put the jacket cuff into the glove barrel, as shown in Fig. 1-12.

Fig. 1-12 Use of insulating gloves

(2) Place of use. Insulating gloves shall be worn when conducting equipment electricity detection, installing and disassembling grounding wires, plugging in and removing the closing power supply fuse (electromagnetic operating mechanism), closing the isolating switch with an insulating rod, closing the circuit breaker and isolating switch through a transmission mechanism, or contacting the equipment housing and frame when HV equipment is grounded, or in other cases where it is possible to contact the charged body.

(3) It is strictly prohibited to use insulating gloves on the operating handle.

(4) Insulating gloves are made of special rubber, and their length shall exceed 10 cm above the wrist. Any ordinary, medical, or chemical gloves cannot be used as a substitute for insulating gloves, nor shall insulating gloves be used for other purposes.

Ⅱ. Insulating boots (shoes)

(Ⅰ) Functions of insulating boots (shoes)

The function of insulating boots (shoes) is to insulate the human body from the ground. Insulating boots (shoes) are safety appliances used to maintain insulation from the ground and prevent electric shock by step voltage when working on the ground of any voltage class(See Fig. 1-13).

(Ⅱ) Inspection of insulating boots (shoes)

The inspection items of insulating boots (as shown in Fig. 1-14) are as follows:

Fig. 1-13　Electric shock by step voltage

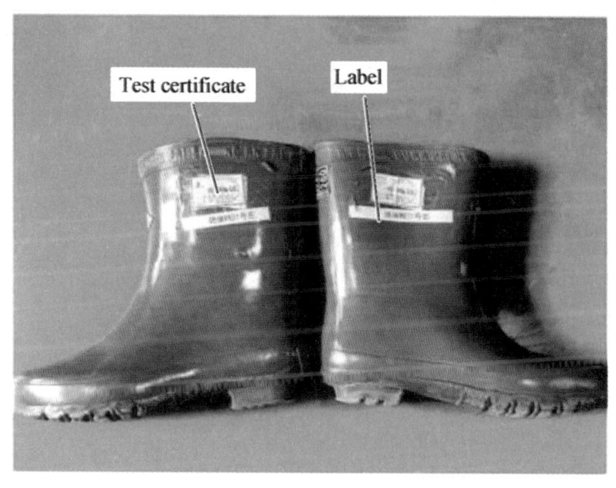

Fig. 1-14　Insulating boots (shoes)

(1) Check whether the label and test certificate are complete before use. The label shall clearly indicate the number of this pair of insulating boots (shoes), which one is for left foot and which one is for right foot, and the shoe size. The date of use shall be within the period of validity of the test, and the test period is half a year.

(2) Before use, the appearance shall be checked for dryness, cleanliness, and no signs of scratches, cracks, broken holes, burrs, bubbles, burns, etc.

(3) The interior of the insulating boots shall not be damp.

(4) Check and ensure that there is no wear on the bottom of the insulating boots, neither yellow adhesive is exposed.

(Ⅲ) Use of insulating boots (shoes)

(1) When using insulating boots, the trouser legs shall be put into the boot barrels, as shown in Fig. 1-15.

(2) Avoid contacting sharp objects, high-temperature or corrosive substance, and prevent insulating boots (shoes) from being damaged.

(3) In case of thunderstorm or primary system grounding, insulating boots should be worn for routine inspection of indoor and outdoor HV equipment. When the resistance of the grounding grid does not meet the requirements, insulating boots should be worn on sunny days.

(4) Insulating boots are made of special rubber. Rubber shoes can not be used instead of insulating boots. It is strictly forbidden to use insulating boots for other purposes.

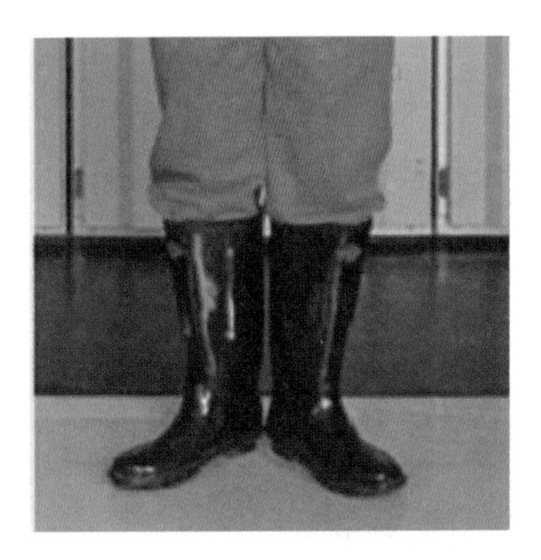

Fig. 1-15 Use of insulating boots (shoes)

任务四　一般防护安全工器具的检查与使用

本任务主要讲安全帽、安全带、升降板、脚扣、安全色、安全标示牌及安全围栏 7 种一般防护安全工器具。通过学习，学习者应掌握其作用、结构、使用及检查方法等。

一、安全帽

（一）安全帽的作用

安全帽具有缓冲减震、分散应力的作用。当物体从高处坠落击向头部时，安全帽可以防止工作人员头部受伤或降低头部受伤害的程度。任何人进入生产现场（办公室、会议室、控制室、值班室和检修班组室除外），必须正确佩戴安全帽。

（二）安全帽的结构

安全帽由帽檐（帽舌）、吸汗带、下颌带、下颌带调节器、帽衬接头、帽箍、后箍、后箍调节器、托带、托带衬垫以及外壳等部位组成，如图 1-16 所示。

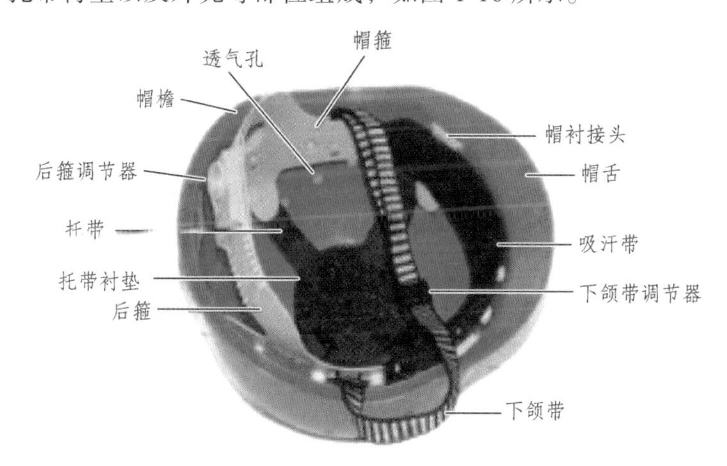

图 1-16　安全帽的结构

（三）安全帽的颜色要求及其使用对象

不同人员配戴的安全帽颜色是有区别的。一般管理或领导人员佩戴红色，运行值班人员佩戴黄色，外来检查以及参观人员佩戴白色，现场作业人员（检修、试验、施工）佩戴蓝色，安全帽示意图如图 1-17 所示。

（四）安全帽的检查

安全帽的检查事项如图 1-18 所示。

图 1-17　安全帽

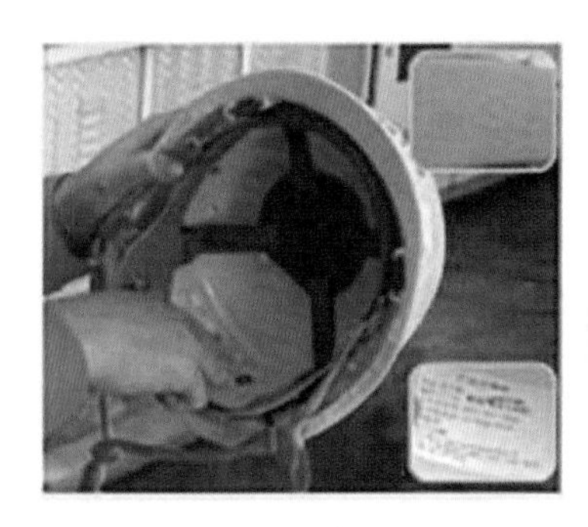

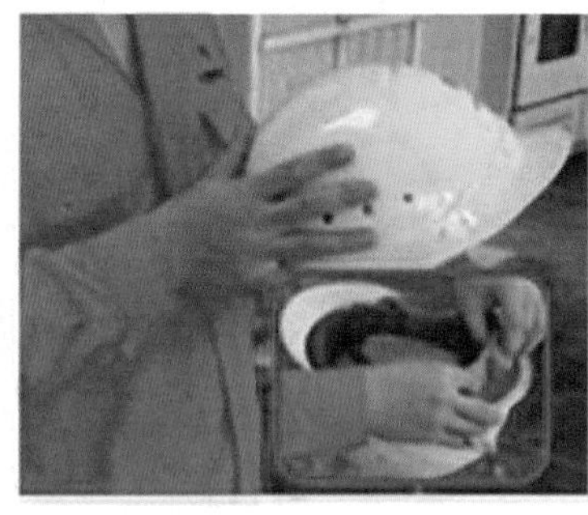

图 1-18　安全帽的检查

（1）合格证、生产日期。检查安全帽的使用周期，从产品制造完成之日起计算，植物枝条编织帽不超过两年；塑料帽、纸胶帽不超过两年半；玻璃钢橡胶帽不超过三年半。对到期的安全帽，应进行抽查测试，合格后方可使用，以后每年抽检一次，抽检不合格，则该批安全帽报废。

（2）外观。使用安全帽前应进行外观检查，检查安全帽的外壳有无龟裂、下凹、裂痕和磨损等情况，发现异常现象要立即更换，不准继续使用。任何受过重击、有龟裂的安全帽，不论有无损坏现象，均应报废。

（3）连接部件。检查并确保安全帽的帽檐、后箍、透气孔、帽衬接头完好，吸汗带无破损，下颌带、顶衬、顶衬托带无破损、断裂，后箍调节器、下颌带调节器能灵活调节，卡位牢固，顶衬与帽顶的距离应在 25 ~ 50 mm，才符合安全和通风要求。

（五）安全帽的使用

（1）戴安全帽前应将后箍按自己头形调整到适合的位置，然后将帽内弹性带系牢。缓冲衬垫的松紧由带子调节，人的头顶和帽体内顶部的空间垂直距离一般在 25 ~ 50 mm，这样才能保证当遭受到冲击时，帽体有足够的空间可缓冲，平时也有利于头和帽体之间的通风。

（2）佩戴时，双手持帽檐，将安全帽从前至后扣于头顶，调整好后箍，系好下颌带。

（3）一定要将安全帽戴正、戴牢，不能晃动，要系牢下颌带，调节好后箍，保持松紧适度，这样不至于被大风吹掉，或者是被其他障碍物碰掉，亦或是由于头的前后摆动使其脱落。

（4）将长头发束好，放入安全帽内，以免低头或弯腰时头发被缠进旋转的机械里。

（5）安全帽体顶部除了在帽体内部安装了帽衬外，有的还开了小孔通风。在使用时不要为了透气而随便再行开孔，从而降低帽体的防护强度。

（6）严禁工作现场出现以下行为：

① 进入生产、施工现场没有正确佩戴安全帽。

② 佩戴不合格的安全帽（质量不合格、颜色不合格）。

③ 佩戴安全帽时，将下颌带放在帽内、脑后或不系紧。

④ 将安全帽当凳子坐。

⑤ 将安全帽乱丢乱放。

⑥ 用安全帽盛装水等物品。

二、安全带

（一）安全带的作用

安全带如图 1-19 所示，是高空作业人员预防高空坠落伤亡事故的防护用具，在高空从事安装、检修、施工等作业时，为预防作业人员从高空坠落，必须使用安全带进行保护。

图 1-19　安全带

（二）安全带的检查

1. 试验日期

检查试验合格证上的试验日期是否在有效期内。安全带每年进行一次静负荷试验。

2. 外观及连接部件

（1）表面整洁、无油污或化学腐蚀。

（2）腰带、围杆带、肩带、腿带、围杆绳、安全绳等带体无灼伤、脆裂及霉变，表面无明显磨损及切口；绳子无断股、无扭结；护腰带接触腰的部分应垫有柔软材料，边缘圆滑无角。

（3）织带折头连接应使用缝线，不应使用铆钉、胶粘、热合等工艺，缝线颜色与织带应有区分。

（4）金属配件表面光洁，无裂纹、无严重锈蚀和可见的变形，金属环类零件不允许使用焊接，不应有开口。

（5）金属挂钩等连接器应有保险装置，应在两个及以上明确的动作下才能打开，且操作灵活；钩体和钩舌的咬口完整，两者不得偏斜，调节装置灵活可靠。

（6）零部件无材料或制造缺陷，塑料件无变形等。

3. 现场拉力试验

使用前，应对围杆带、安全绳做拉力试验，如图 1-20（a）所示，拉力试验后应检查连接受力位置是否有撕裂、破损。

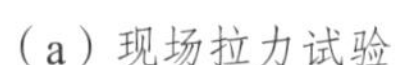

（a）现场拉力试验

（b）高挂低用

图 1-20　安全带的检查和使用

（三）安全带的使用

安全带使用时务必系好挂牢、做好保护工作。

（1）使用场合。在没有脚手架或者在没有栏杆的脚手架上工作，高度超过 1.5 m 时，应使用安全带；凡在离地面 2 m 及以上的地点进行工作，应使用双保险安全带，或采用其他可靠的安全措施；当使用 3 m 以上的安全绳时，应配合缓冲器使用。

（2）高挂低用。安全带的使用要遵循高挂低用的原则，如图 1-20（b）所示，禁止低挂高用。

（3）扎在结实牢固的构件上。安全带的受力点宜在腰部与臀部之间位置，安全带的挂钩或绳子应挂在结实牢固的构件上，禁止挂在移动或不牢固的物件上。

（4）转位时不得失去安全带保护。穿戴好后应仔细检查连接扣或调节扣，确保各处绳扣连接牢固。使用中应防止摆动、碰撞，避开尖刺、高温及明火。在杆塔上转位时，不得失去安全带保护。

（5）安全带与系带不能打结使用，安全绳应是整根，不得私自接长使用，也不应打结，应通过连接器连接。

三、升降板

（一）升降板的作用

升降板是攀登水泥电杆的主要工具之一。其优点是适应性强，工作方便。不论电杆直径大小是否有变化，升降板均适用，且方便高空作业人员站立，缓解工作疲劳。

（二）升降板的结构

升降板由两条脚踏板、吊绳（踏板绳）、金属挂钩组成，如图 1-21 所示。

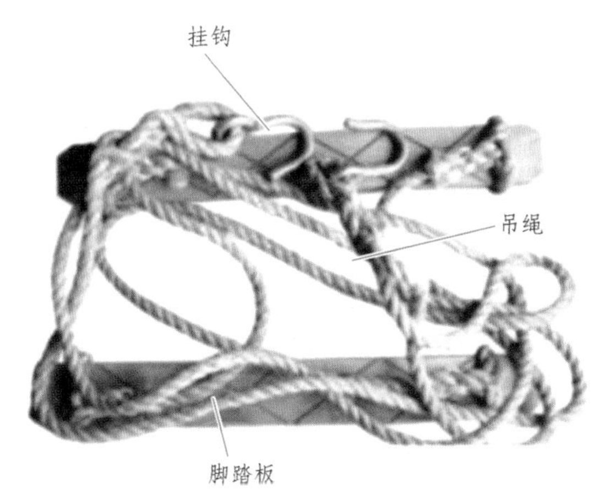

图 1-21　升降板的结构

（1）脚踏板：由硬质木材制作，一般长 630 mm、宽 75 mm、厚 25 mm，表面刻有防滑斜纹。
（2）吊绳：用直径为 16 mm 的优质棕绳制作，呈三角形状绑扣。
（3）挂钩：由直径不小于 10 mm 的镀锌圆钢制作。

（三）升降板的检查

（1）每半年定期进行试验一次。
（2）确保连接部位牢固，绑扎结实可靠。
（3）确保绳索无扭结、脱股、断股。
（4）确保脚踏板结实，木质无腐蚀、劈裂。
（5）确保金属挂钩无损伤及变形等。
（6）确保金属绑扎线组件无毛刺、损伤及断股。
（7）定期检查并有记录，未超期使用。

（四）升降板的使用

先将升降板系在电杆上离地 0.5 m 处，人站在板上用力向下踩蹬，测试绳索不断股、踏板不折裂方可使用。使用升降板时，要保持人体平稳不摇晃且保持站立姿势。升降板不能随意从杆上往下扔，以免摔坏。

四、脚扣

（一）脚扣的作用

脚扣是攀登水泥杆的主要用具之一，脚扣的半圆环和根部装有橡胶套，可用来防滑。

（二）脚扣的结构

脚扣由脚踏板、小爪、橡胶防滑块、套管、皮带、金属母材围杆钩 6 个部件构成，如图 1-22 所示。

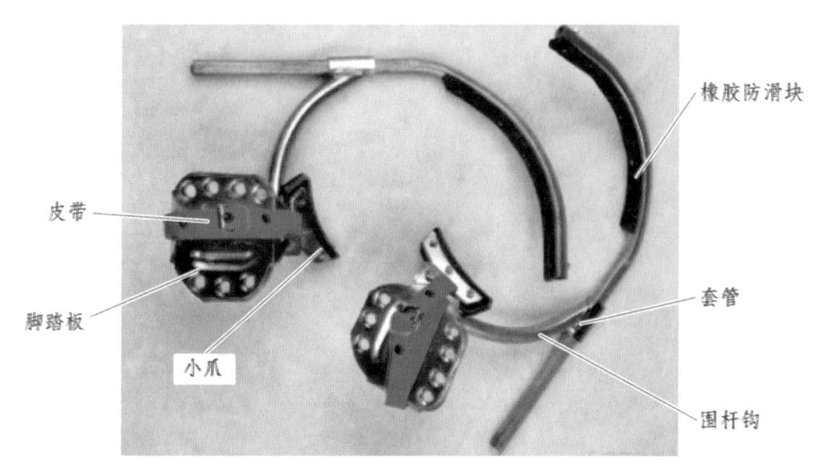

图 1-22 脚扣的结构

（三）脚扣的检查及使用（图 1-23）

（1）检查试验日期。登杆前，按电杆的直径选择脚扣大小，检查试验合格证的试验日期，应保证在有效期内。每年进行一次预防性试验，每月进行一次外观检查。

（a）检查试验日期

（b）检查外观及连接部件

（c）做人体冲击试验

（d）检查杆根、杆身

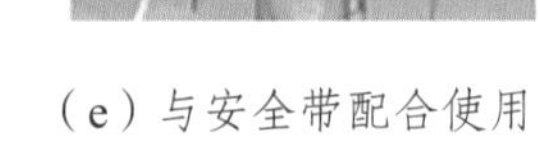

（e）与安全带配合使用

（f）调整脚扣大小

图 1-23 脚扣的检查及使用

（2）检查外观及连接部件。使用前应做外观检查，检查各部位应无裂纹、腐蚀、开焊、变形等现象；橡胶防滑块（套）完好，无破损；皮带完好，无霉变、裂缝或严重变形；小爪连接牢固，活动灵活。

（3）做人体冲击试验。登杆前应对脚扣做人体冲击试验，方法是将脚扣系于电杆离地 0.5 m 左右处，借人体重量猛力向下蹬踩，确认脚扣及胶套没有变形及任何损坏后方可使用。

（4）检查杆根、杆身。登杆前，应检查杆根牢固，杆身无裂纹。

（5）必须与安全带配合使用。登杆时，必须与安全带配合使用，以防登杆过程中发生坠落事故。

（6）调整脚扣大小。登杆前应将脚扣登板的皮带系牢，登杆过程中，应根据杆径的改变相应调整脚扣大小。

脚扣虽是攀登电杆的防护安全工器具，但作业人员应经过较长时间的练习、熟练地掌握使用方法后，才能让其起到防护作用；若使用不当，也会发生人身伤亡事故。不准随意将脚扣从杆上向下摔扔，作业前后应轻拿轻放，并妥善保存在工具柜内。

五、安全色

（一）定义

安全色是表达安全信息的颜色，表示禁止、警告、指令、提示等意义。

（二）特点（见表 1-1）

表 1-1　安全色的特点

颜色	特点	含义	应用实例
红色	注目性非常高，视认性很好	禁止、停止、消防和危险	禁止合闸 有人工作
黄色	在太阳光直射下颜色较明显	警告、注意	止步 高压危险
蓝色	对人眼能产生比红色还高的明亮度	指令、必须遵守的规定	必须戴安全帽

续表

颜色	特点	含义	应用实例
绿色	能使人联想到大自然的一片翠绿	提示、安全状态通行	紧急出口 EXIT

（三）使用场合

安全色用途广泛，如用于安全标示牌、交通标示牌、防护栏杆及机器上不准乱动的部位、紧急停止按钮、安全帽、吊车、升降机、行车道中线等。

六、安全标示牌

安全标示牌主要设置在容易发生事故或危险性较大的工作场所，主要分为禁止标示牌、警告标示牌、指令标示牌、提示标示牌及其他标示牌。

禁止标示牌如“禁止合闸，有人工作”“禁止合闸，线路有人工作”“禁止攀登，高压危险”等；警告标示牌如“止步，高压危险!”“当心触电”等；指令标示牌如“必须戴安全帽”“必须系安全带”等；提示标示牌如“从此进出”“从此上下”等。

（一）几种常用的标示牌（见图 1-24）

图 1-24　几种常用的标示牌

1. “禁止合闸有人工作”标示牌

（1）设置在一经合闸即可送电到已停电检修（施工）设备的开关和刀闸的操作把手上。

（2）设置在已停电检修（施工）设备的电源开关或合闸按钮上。
（3）设置在显示屏上进行操作的断路器（开关）和隔离开关（刀闸）的操作处。

2.“禁止攀登高压危险”标示牌

（1）设置在架空输电线路杆塔脚钉或爬梯侧。
（2）设置在台架变压器上，可挂于主、副杆上及槽钢底的行人易见位置，也可使用支架安装。
（3）设置在户外电缆保护管或电缆支架上。
（4）设置在标示牌底边距地面 2.5 ~ 3.5 m。

3.“禁止合闸线路有人工作”标示牌

当线路上有人工作：
（1）设置在已停电检修（施工）的电力线路的开关和刀闸的操作把手上。
（2）设置在显示屏上进行操作的断路器（开关）和隔离开关（刀闸）的操作处。

4.“止步高压危险”标示牌

（1）设置在应装设的临时遮栏（围栏）上。
（2）高压开关柜内手车开关拉出后，隔离带电部位的挡板封闭后禁止开启，并设置“止步，高压危险”的标示牌。
（3）在室外构架上工作，则应在工作地点邻近带电部分的横梁上，悬挂“止步，高压危险”的标示牌。
（4）设置在配电房的正门及箱式变压器、电缆分接箱的外壳四周。
（5）设置在台架变压器、坐地式台式变压器的围墙、围栏及门上。
（6）设置在户内变压器的围栏或变压器室门上。

5.“在此工作”标示牌

设置在工作地点或检修设备上。

6.“从此上下”标示牌

设置在现场工作人员可以上下的脚架、爬梯上。

7.“从此进出”标示牌

设置在围栏的出入口处。

（二）其他标示牌

除以上常用的安全标示牌外，电力生产中还有很多其他标示牌，如图 1-25 所示。

注意通风

必须戴安全帽

必须戴防护手套

必须系安全带

必须穿防护鞋

必须戴防毒面具

必须戴防护镜

必须拔出插头

止步 高压危险

有电 高压危险

有电 注意安全

当心触点

图 1-25 其他标示牌

七、安全围栏

安全围栏如图 1-26 所示。

图 1-26 安全围栏

在部分停电的设备上工作，或安全距离小于设备不停电的安全距离的未停电设备上工作，应装设临时安全围栏，临时安全围栏与带电部分的距离不得小于作业人员工作中正常活动范围与设备带电部分的安全距离，临时安全围栏可由干燥木材、橡胶或其他坚韧绝缘材料制成，装设应牢固，并悬挂“止步，高压危险!”的标示牌。

在室外高压设备上工作，应在工作地点四周装设安全围栏，其出入口要围至临近道路旁边，并设有“从此进出”的标示牌。工作地点四周围栏上面向带电侧悬挂适当数量的“止步，高压危险!”标示牌，标示牌应朝向围栏里面。若室外配电装置的大部分设备停电，只有个别地点保留有带电设备而其他设备无触及带电导体的可能时，可以在带电设备四周装设全封闭围栏，围栏上悬挂适当数量的“止步，高压危险!”标示牌，标示牌应朝向围栏外面。

禁止翻越围栏。禁止作业人员擅自移动或拆除围栏、标示牌。因工作原因必须短时移动或拆除围栏、标示牌，应征得工作许可人同意，并在工作负责人的监护下进行。工作完毕后立即恢复。

直流换流站单极停电工作，应在双极公共区域设备与停电区域之间设置围栏，在围栏面向停电设备及运行阀厅门口悬挂“止步，高压危险!”标示牌。在检修阀厅和直流场设备处设置“在此工作”的标示牌。

Task Ⅳ Inspection and use of general protection safety tools and instruments

In this module, 7 types of general protection safety tools and instruments, including safety helmets, safety belts, lifting boards, grapplers, safety colors, safety sign boards, and safety fences, are introduced. Master their functions, structures, uses, and inspection methods through study.

Ⅰ. Safety helmets

(Ⅰ) Functions of safety helmet

A safety helmet has the function of cushioning, shock absorption, and stress dispersion. When an object falls from a height and hits the head, the safety helmet can prevent the worker from being injured and can reduce the degree of head injury. Anyone entering the production site (except offices, conference rooms, control rooms, duty rooms and maintenance rooms) must wear safety helmets correctly.

(Ⅱ) Structure of safety helmet

The safety helmet consists of a brim (visor), a sweat absorbing belt, a jaw belt, a jaw belt regulator, a cap liner joint, a hat hoop, a rear hoop regulator, a bracket, a bracket liner and an outer shell, as shown in Fig. 1-16.

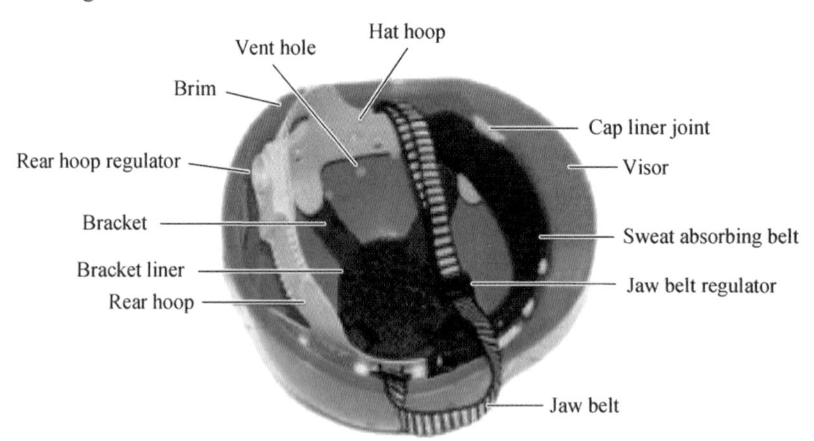

Fig. 1-16 Structure of safety helmet

(Ⅲ) Color requirements and use objects for safety helmets

The color of safety helmets varies from different jobs. General managers or leaders wear red safety helmets, on-duty operators wear yellow safety helmets, external inspectors and visitors wear white safety helmets, and on-site operators (maintenance, testing, construction) wear blue safety helmets. Safety helmets is shown in Fig. 1-17.

Fig. 1-17　Safety helmet

(Ⅳ) Inspection of safety helmet

The inspection items of the safety helmet are as shown in Fig. 1-18.

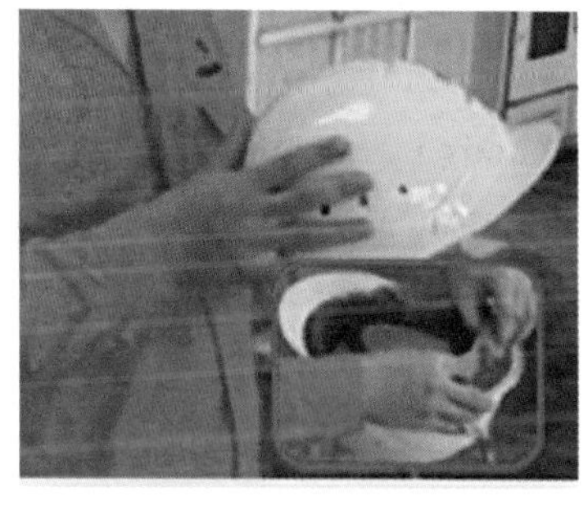
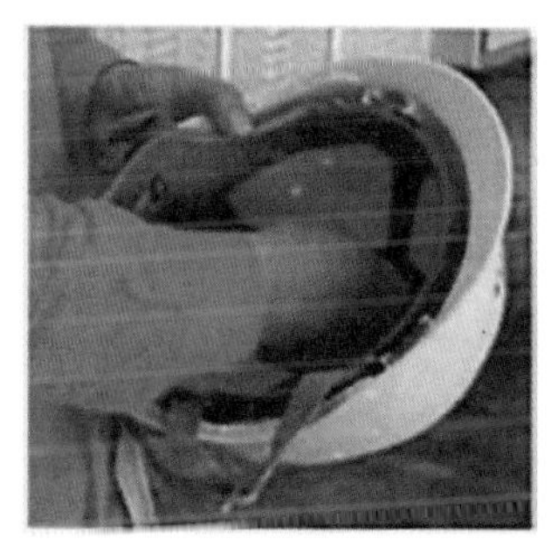

Fig. 1-18　Inspection of safety helmet

(1) Certificate of conformity, production date. Check the service life of the safety helmet, which shall be calculated from the date of product manufacturing completion. The service life of plant branch braided helmet shall not exceed two years. The service life of plastic helmet and paper rubber helmet shall not exceed two and a half years. The service life of fiberglass rubber helmet shall not exceed three and a half years. For the expired safety helmets, spot checks shall be carried out and they can be used only after they are qualified. In the future, a random inspection will be conducted once a year. If they fail the random inspection, the batch of safety helmets will be scrapped.

(2) Appearance. Visual inspection shall be carried out before using the safety helmet to check the safety helmet shell for chapping, depressions, cracks and wear, and other conditions. If any abnormalities are found, replace them immediately, and do not continue to use them. Any hard-hit, cracked safety helmet, regardless of whether it is damaged or not, shall be scrapped.

(3) Connecting component. Check the brim, rear hoop, and vent hole of the safety helmet. The cap liner joint is intact, the sweat absorbing belt is not damaged, the jaw band, the top lining, and the top lining bracket are not damaged or broken, the rear hoop regulator and the jaw band regulator can be adjusted flexibly, the clamping position is firm, and the distance between the top liner and the top shall be between 25-50 mm in order to meet the requirements of safety and ventilation.

(Ⅴ) Use of safety helmet

(1) Before wearing a safety helmet, the rear hoop shall be adjusted to a suitable position according to its own head shape, and then the elastic belt inside the helmet shall be fastened. The tightness of the buffering cushion is adjusted by the belt. The vertical distance between the top of a person's head and the top of the helmet body is generally between 25 and 50 mm, which ensures that the helmet body has sufficient space to cushion when subjected to impact and is also conducive to ventilation between the head and the helmet body.

(2) When wearing, hold the brim of the helmet in both hands, buckle the safety helmet from front to back on the top of head, adjust the rear hoop, and fasten the jaw belt.

(3) Be sure to wear the helmet straight and firmly, do not shake it. Fasten the jaw belt firmly, adjust the rear hoop and keep it moderately loose, so that it won't be blown off by strong winds, hit by other obstacles, or the safety helmet may fall off due to the head swinging back and forth.

(4) Tie up the long hair and put it in safety helmet to avoid getting tangled up in the rotating machine when bowing head or bending over.

(5) The top of the safety helmet body is not only equipped with a liner inside the body, but also has a small hole for ventilation. When in use, do not open holes casually, which can cause the reduction of the protective strength of the helmet body.

(6) The following behaviors at the work site are strictly prohibited:

① Anyone who enters the production and construction site does not wear a safety helmet correctly.

② Wear unqualified safety helmets (unqualified quality, unqualified color).

③ When wearing a safety helmet, place the jaw belt in the helmet, without tying it tightly behind the head.

④ Sit on the safety helmet as a stool.

⑤ Throw the safety helmet around.

⑥ Use a safety helmet to hold water and other items.

Ⅱ. Safety belt

(Ⅰ) Functions of a safety belt

As shown in Fig. 1-19, safety belts are protective equipment for high-altitude operators to prevent casualties caused by falling from high altitude. In order to prevent operators from falling from height, safety belts must be used to protect them when they are engaged in installation, maintenance, construction and other operations at high altitude.

Fig. 1-19 Safety belt

(Ⅱ) Inspection of safety belt

1. Test date

Check whether the test date on the test certificate is within the period of validity. The safety belt shall undergo a static load test once a year.

2. Appearance and connecting component

(1) The surface is clean and free from oil stains or chemical corrosion.

(2) Belts, girdles, shoulder straps, leg straps, girder ropes, safety ropes, etc., have no burns, brittleness and mildew, and no obvious wear and tear on the surface. The rope has no broken strands and no kink. The part where the belt touches the waist shall be padded with soft material, and the edges shall be smooth without sharp points.

(3) Stitches shall be used for ribbon folding connections, rivets, adhesives, heat bonding and other processes shall not be used. Stitch colors shall be distinguished from ribbons.

(4) The surface of metal fittings shall be smooth and clean, without cracks, severe rust, and visible deformation. Metal ring parts are not allowed to be welded and shall not have openings.

(5) Metal hooks and other connectors shall have safety devices that can only be opened with two or more clear actions, and the operation shall be flexible. The seam of the hook body and the hook tongue shall be complete, both must not be deflected, and the adjusting device shall be flexible and reliable.

(6) The parts and components shall have no material or manufacturing defects, and the plastic parts shall have no deformation, etc.

3. On-site tension test

Before use, a tension test shall be conducted on the girdle and safety rope, as shown in Fig. 1-20(a). After the tension test, the connection shall be checked for tearing or damage at the stress position.

(a) On-site tension test

(b) Hanging high and using low

Fig. 1-20 Inspection and use of safety belt

(Ⅲ) Use of safety belt

When using the safety belt, be sure to fasten and hang it securely without losing protection.

(1) Place of use. When working without scaffold or on scaffolding without handrails, safety belts shall be used when the height exceeds 1.5 m. When working at a height of 2 m or more, double safety belts or other reliable safety measures shall be used. When using more than 3 m safety rope, it shall be used with the buffer.

(2) Hanging high and using low. The use of safety belt shall follow the principle of hanging high and using low, as shown in Fig. 1-20 (b). It is prohibited to hang low and use high.

(3) Tied to a sturdy and firm member. The stress point of the safety belt should be located between the waist and the buttocks, and the hook or rope of the safety belt shall be hung on sturdy and firm members, and it is forbidden to hang on moving or unstable objects.

(4) Do not lose safety belt protection during position transfer. After wearing, carefully check the connecting buckle or adjusting buckle to ensure that the buckle is firmly connected everywhere. In use, swing and collision shall be prevented, sharp thorns, high temperature and open fire shall be avoided. When transferring position on the tower, the safety belt protection must not be lost.

(5) Safety belts and straps cannot be used in knots. The safety belt shall be a whole piece and shall not be extended or tied without authorization. It shall be connected through connectors.

Ⅲ. Lifting board

(Ⅰ) Functions of lifting board

The lifting board is one of the main tools for climbing cement poles. Its advantages include strong adaptability and convenient work. No matter whether the diameter of the pole changes or not, the lifting board is applicable, and it is convenient for high-altitude operators to stand and alleviate work fatigue.

(Ⅱ) Structure of lifting board

The lifting board consists of two foot pedals, lifting rope (pedal rope), and metal hook, as shown in Fig. 1-21.

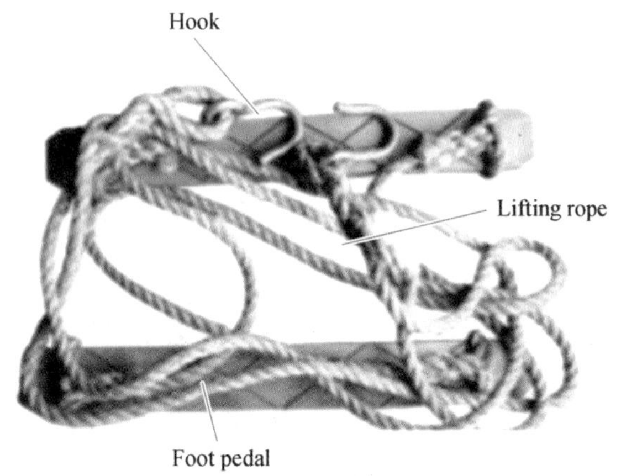

Fig. 1-21 Structure of lifting board

(1) Foot pedal: Made of hardwood, generally 630 mm long, 75 mm wide, and 25 mm thick, with anti-skid twill engraved on the surface.

(2) Lifting rope: Made of high-quality brown rope with a diameter of 16 mm, fastened in a triangular shape.

(3) Hook: Made of galvanized round steel with a diameter of not less than 10 mm.

(Ⅲ) Inspection of lifting board

(1) Conduct regular test every six months.

(2) The connection part shall be firm and the binding shall be firm and reliable.

(3) The rope shall be free from twisting, detachment, or breakage.

(4) The foot pedal shall be strong, and the wood is non-corrosive and not split.

(5) The metal hook has no damage and deformation.

(6) The metal binding thread assembly is free of burrs, damage and broken strands.

(7) Regularly inspect and keep records, and do not exceed the usage period.

(Ⅳ) Use of lifting board

First, tie the lifting board on the pole 0.5 m from the ground, a person stands on the board and pedals down vigorously. If the test rope is continuous and the pedal does not break, the lifting board can be used. When using the lifting board, keep the body steady and keep standing posture. The lifting board cannot be thrown down from the pole at will to avoid damage.

Ⅳ. Grappler

(Ⅰ) Functions of grappler

The grappler is one of the main tools for climbing the cement pole. The semicircle ring and the root of the grappler buckle are equipped with a rubber sleeve, which can be used for skid resistance.

(Ⅱ) Structure of grappler

The grappler is composed of six components: foot pedal, small claw, rubber anti-skid block, bushing, belt and metal base metal rod hook, as shown in Fig. 1-22.

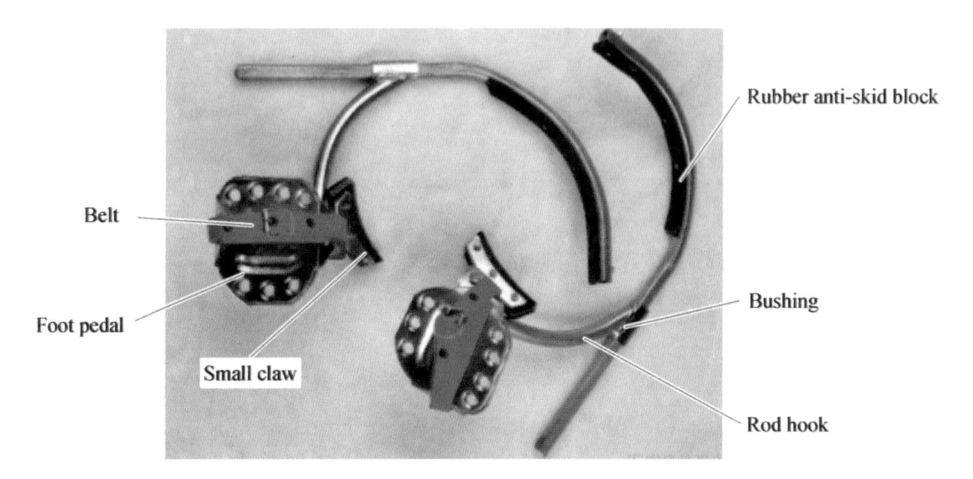

Fig. 1-22 Structure of grappler

(Ⅲ) Inspection and use of grappler (see Fig. 1-23)

(1) Check the test date. Before climbing the pole, select the grappler size according to the diameter of the pole, and check whether the test date of the test certificate is within the period of validity. Conduct preventive test once a year and conduct visual inspection once a month.

(2) Check appearance and connecting component. Before use, visual inspection shall be done to check that all parts shall be free of cracks, corrosion, welding, deformation and so on. The rubber anti-skid block (sleeve) is intact and undamaged; The belt is in good condition without mildew, crack or serious deformation; The small claws are firmly connected and flexible in movement.

(3) Conduct a human impact test. Before climbing the pole, a human impact test shall be conducted on the grappler. The method is to tie the grappler to the pole about 0.5 m above the ground, and use the weight of the human body to forcefully push it down. Only after confirming that the grappler and rubber sleeve are not deformed or damaged can they be used.

(4) Check the pole footing and body. Before climbing the pole, check that the pole footing is firm and there are no cracks in the pole body.

(5) It must be used in conjunction with safety belt. When climbing, it must be used in conjunction with safety belt to prevent falling accidents during the climbing process.

(6) Adjust the grappler size. Before climbing the pole, the belt of the grappler board shall be securely fastened. During the climbing process, the size of the grappler shall be adjusted accordingly according to the change in the pole diameter.

(a) Check the test date

(b) Check appearance and connecting component

(c) Conduct a human impact test

(d) Check the pole footing and body

(e) Use in conjunction with safety belt

(f) Adjust the grappler size

Fig. 1-23 Inspection and use of grappler

Although the grapplers are protective safety tools and instruments for climbing poles, the operators shall practice for a long time and skillfully master the method of using them before they can play a protective role. If it is not used properly, personal casualties will also occur. Do not throw the grappler off the pole at will. Handle it with care before and after work, and store it properly in a tool cabinet.

Ⅴ. Safety colors

(Ⅰ) Definition

Safety colors are colors that express safety information, indicating meanings such as prohibition, warning, command, and prompt.

(Ⅱ) Characteristics (Tab. 1-1)

Tab. 1-1 Characteristics of safety colors

Color	Characteristics	Meaning	Application example
Red	Very eye-catching, with good visual recognition	Prohibition, stop, fire fighting, and hazard	 No switch closing! Working!
Yellow	More obvious in direct sunlight	Warning and attention	 Stop! High voltage, danger!
Blue	Can produce a higher brightness than red for the human eye	Command and mandatory regulations	 Must wear safety helmet
Green	Can remind people of a verdant green of nature	Prompt, safe state passage	 紧急出口 EXIT

(Ⅲ) Place of use

Safety color has a wide range of uses, such as safety sign boards, traffic sign boards, protective handrails and parts on the machine that are not allowed to move, emergency stop buttons, safety helmets, cranes, lifts, the middle line of the carriageway and so on.

Ⅵ. Safety sign boards

Safety sign boards are mainly located in accident-prone or dangerous workplaces, which are mainly divided into forbidding sign boards, warning sign boards, command sign boards, prompt sign boards, and other sign boards.

Forbidding sign boards include "No switch closing! Working! ", "No switch closing! Working on the line!", "Do not climb! High voltage, danger!", etc. Warning sign boards include "Stop! High voltage, danger!" "Danger! Electric shock!". Command sign boards include "Must wear safety helmet", "Must fasten safety belt". Prompt sign boards include "Enter and exit from here", "Get up and down from here".

(I) Several commonly used sign boards (see Fig. 1-24)

Fig. 1-24 Several commonly used sign boards

1. "No switch closing! Working! " sign board

(1) Set up on the operating handle of the switch and knife switch which can send power to the equipment with interruption maintenance (construction) as soon as the switch is closed.

(2) Set up on the power supply switch or closing button of the equipment with interruption maintenance (construction).

(3) Set up on the operating place of the circuit breaker (switch) and the isolating switch (knife switch) operated on the display screen.

2. "Do not climb! High voltage, danger!" sign board

(1) Set up on the foot nail or climbing side of the tower of the overhead transmission line.

(2) Set up on the bench transformer and can be hung on the main rod, auxiliary rod and the pedestrian visible position at the bottom of the groove steel, or it can be installed with a support.

(3) Set up on outdoor cable protection pipes or cable supports.

(4) Set at the point where the bottom edge of the sign board is 2.5 ~ 3.5 m above the ground.

3. "No switch closing! Working on the line!" sign board

When someone is working on the line:

(1) Set up on the operating handle of the switch and knife switch of the electric power line with interruption maintenance (construction).

(2) Set up on the operating place of the circuit breaker (switch) and the isolating switch (knife switch) operated on the display screen.

4. "Stop! High voltage, danger!" sign board

(1) Set up on the temporary barrier (fence) that shall be installed.

(2) After the handcart switch in the HV switch cabinet is pulled out, the baffle plate used to isolate the live part, once closed, shall not be opened, and a sign board reading "Stop! High voltage, danger!" shall be provided.

(3) When working on an outdoor framework, a sign board reading "Stop! High voltage, danger!" should be hung on the cross beam near the live part on the work site.

(4) Set up around the main door of the distribution room and the housing of the box type transformer and cable distribution box.

(5) Set up on the enclosing walls, fences, and doors of bench transformer and floor mounted bench transformer.

(6) Set up on the fence of the indoor transformer or the transformer room door.

5. "Work here" sign board

Set up at the workplace or equipment under maintenance.

6. "Get up and down from here" sign board

Set up on the foot stool and climbing ladder where the worker can get up and down.

7. "Enter and exit from here" sign board

Set at the entrance and exit of the fence.

(Ⅱ) Other sign boards

In addition to the above commonly used safety sign boards, there are many other sign boards at the power production site, as shown in Fig. 1-25.

Caution ventilation

Must wear safety helmet

Wear must protective gloves

Must fasten safety belt

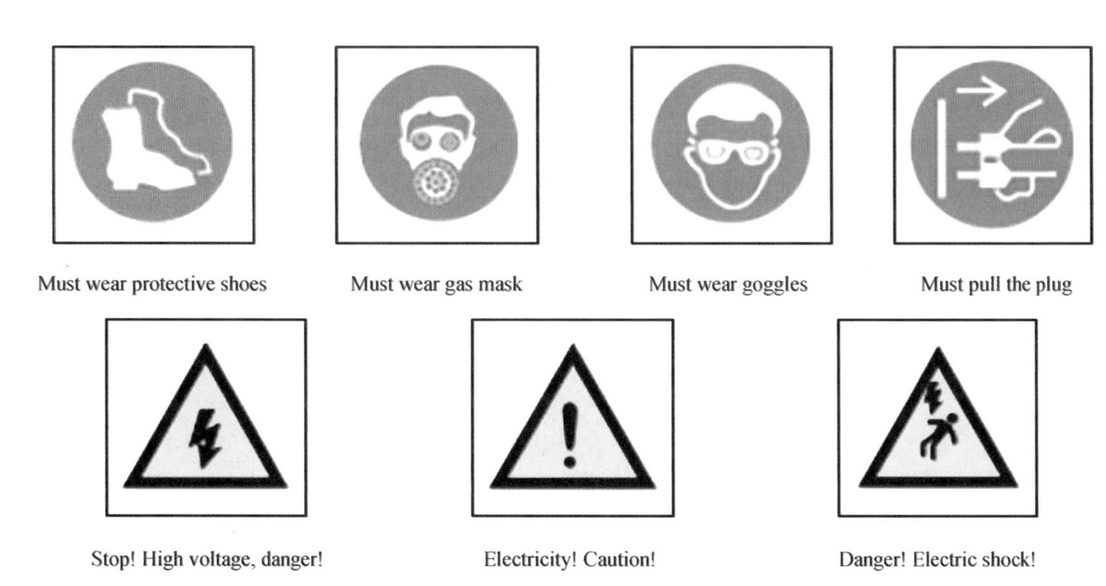

Fig. 1-25 Other sign boards

Ⅶ. Safety fence

The safety fence is as shown in Fig. 1-26.

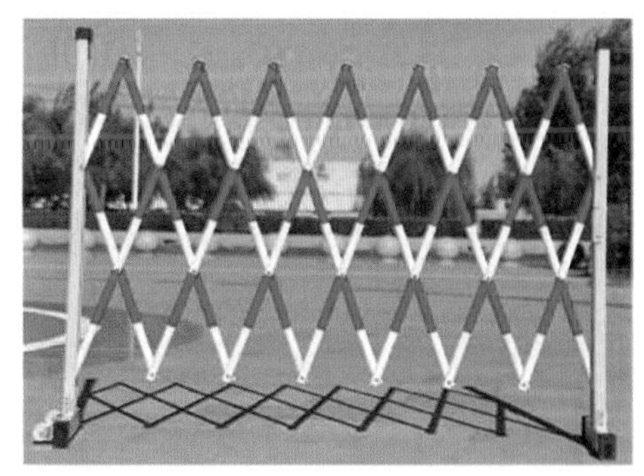

Fig. 1-26 Safety fence

Temporary safety fences shall be installed when working on partially powered-off equipment or on non-powered-off equipment with a safe distance less than the safe distance for non-powered-off equipment. The distance between the temporary safety fence and the live part shall not be less than the safe distance between the normal range of activities of the operator and the live part of the equipment. Temporary safety fences can be made of dry wood, rubber, or other tough and insulating materials, and shall be securely installed and hung with a sign board reading "Stop! High voltage, danger!"

When working on outdoor HV equipment, a fence shall be installed around the work site, with its entrance arranged next to an adjacent road, and a sign board reading "Enter and exit from here" shall be provided. An appropriate number of sign boards reading "Stop! High voltage, danger!" should be hung on the fences towards the live side surrounding the work site, and the sign boards shall face what's inside the fences. If most of the equipment of the outdoor power distribution unit is power-interrupted, live equipment only exists in a few places, and other equipment is not likely to come in contact with the live conductors, a fully enclosed fence can be installed around the live equipment, an appropriate number of sign boards reading “Stop! High voltage, danger!” should be hung on the fence, and the sign boards shall face what's inside the fence.

No crossing. Operators are prohibited from moving or dismantling fences and sign boards without authorization. If it is necessary to move or remove fences and sign boards for a short time due to work reasons, the permission of the work permitter shall be obtained, and the fences and sign boards shall be moved or removed under the supervision of the person in charge of work and shall be re-installed immediately after the work is done.

When working at a DC convertor station with monopole power interruption, a fence shall be provided between the equipment in the bipole common area and the power interruption area, and a sign board reading “Stop! High voltage, danger!” should be hung on the fence (the sign board shall face the power-interrupted equipment) and at the gate of the operating valve hall. Set up “Work here” sign boards at the maintenance valve hall and DC field equipment.

任务五　电力安全工器具的保管与存放

电力安全工器具的保管、存放及报废，必须满足国家和行业标准及产品说明书要求。

一、电力安全工器具的保管与存放

（一）橡胶塑料类电力安全工器具

（1）橡胶塑料类电力安全工器具应存放在干燥、通风、避光的环境下，存放时离开地面和墙壁 20 cm 以上，离开发热源 1 m 以上，避免阳光、灯光或其他光源直射，避免雨雪浸淋，防止挤压、折叠和尖锐物体碰撞，严禁与油、酸、碱或其他腐蚀性物品存放在一起。

（2）防护眼镜保管于干净、不易碰撞的地方。

（3）防毒面具应存放在干燥、通风，无酸、碱、溶剂等物质的库房内，严禁重压。防毒面具的滤毒罐（盒）的贮存期为 5 年（3 年），过期产品应经检验合格后方可使用。

（4）空气呼吸器在贮存时应装入包装箱内，避免长时间暴晒，不能与油酸、碱或其他有害物质共同贮存，严禁重压。

（5）防电弧服贮存前必须洗净晾干，不得与有腐蚀性物品放在一起，应存放于干燥通风处，避免长时间接触地气受潮，防止紫外线长时间照射。长时间保存时，应注意定期晾晒，以免霉变、虫蛀以及滋生细菌。

（6）橡胶和塑料制成的耐酸服存放时应注意避免接触高温，用后清洗晾干，避免暴晒。长期保存应撒上滑石粉，以防粘连。合成纤维类耐酸服不宜用热水洗涤熨烫，避免接触明火。

（7）绝缘手套使用后应擦净、晾干，保持干燥、清洁，最好撒上滑石粉以防粘连。绝缘手套应存放在干燥、阴凉的专用柜内，与其他工具分开成双定置摆放，如图 1-27 所示。其上不得堆压任何物件，以免刺破手套。绝缘手套不允许放在过冷、过热、阳光直射和有酸、碱、药品的地方，以防胶质老化，降低绝缘性能。

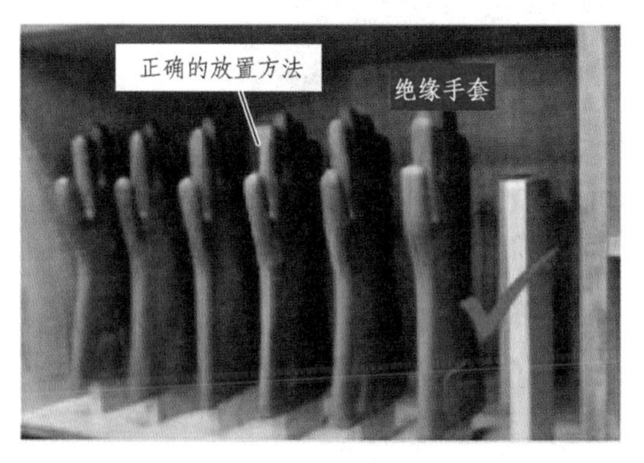

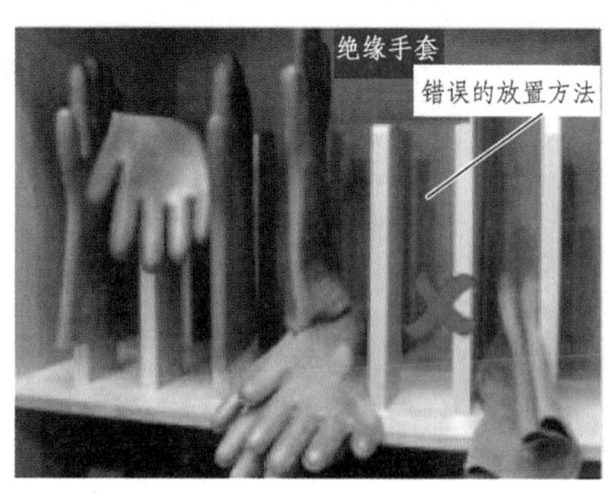

图 1-27　绝缘手套的存放

（8）绝缘靴（鞋）使用后应擦拭干净，不允许放在过冷、过热、阳光直射和有酸、碱、油品、化学药品的地方，应存放在干燥、阴凉的专用柜内或支架上，成双定置摆放，如图 1-28 所示。

图 1-28　绝缘靴的存放

（9）耐酸靴穿用后，应立即用水冲洗，存放阴凉处，撒上滑石粉，以防粘连，应避免接触油类、有机溶剂和锐利物。

（10）防静电鞋和导电鞋应保持清洁。如表面附着污染尘土、油蜡、粘贴绝缘物或因老化形成绝缘层，会大大影响其电阻。刷洗时要用软毛刷、软布蘸酒精或不含酸、碱的中性洗涤剂进行擦洗。

（11）绝缘遮蔽罩使用后应擦拭干净，装入包装袋内，放置于清洁、干燥通风的架子或专用柜内，上面不得堆压任何物件。

（二）环氧树脂类电力安全工器具

环氧树脂类电力安全工器具应置于通风良好、清洁干燥、无阳光直晒、无腐蚀和无有害物质的场所保存，如图 1-29 所示。

（1）绝缘杆使用后必须擦干净，存放时绝缘杆应架在支架上或悬挂起来，不得直接接触地面、墙面，防止受潮、脏污，且应成套放置。

图 1-29　环氧树脂类电力安全工器具的存放

（2）绝缘隔板应统一编号，存放在室内干燥通风、离地面 20 cm 以上专用的工具架上或柜内。如果表面有轻度擦伤，应涂绝缘漆处理。

（3）接地线不用时将软铜线盘好。存放在干燥的室内，宜存放在专用架上，架上的编号与接地线的编号应一致。

（4）核相器应存放在干燥通风的专用支架上或者专用包装盒内。

（5）验电器使用后应存放在防潮盒或绝缘安全工器具存放柜内，置于通风干燥处，见图1-30。

图 1-30　验电器的存放

（6）绝缘夹钳应保存在专用的箱子或柜子里，以防受潮和磨损。

（三）纤维类电力安全工器具

纤维类电力安全工器具应放在干燥、通风、避免阳光直晒、无腐蚀及有害物质的场所，且与热源保持 1 m 以上的距离。

（1）安全带不使用时，应由专人保管。存放时，不应接触高温、明火、强酸、强碱或尖锐物体，不应存放在潮湿的地方。储存时，应对安全带定期进行外观检查，发现异常必须立即更换，检查频次应根据安全带的使用频率确定。

（2）安全绳每次使用后应检查，并定期清洗。

二、电力安全工器具的管理

电力安全工器具应存放在专用的安全工器具室内，建立电力安全工器具管理台账，并抄报公司安监部。应做到账、卡、物相符，试验报告、试验合格证、检查记录齐全。

电力安全工器具应由专人保管，并做好领用记录。保管人应定期进行日常检查、维护、保养，发现不合格或超试验周期的应另外存放，标记不准使用的标志，停止使用，并及时送公司安监部统一报废。各单位负责人、兼（专）职安全员应按规定组织对电力安全工器具使用前的检查，并做好检查记录。电力安全工器具均有专门的用途和使用范围，必须正确使用、妥善保管，严禁挪作他用。

要对作业人员进行安全培训，严格执行操作规定，正确使用电力安全工器具。不熟悉使用操作方法的人员不得使用电力安全工器具。

各单位对电力安全工器具的正确使用应严格监督管理，对不按规定正确佩戴和使用电力安全工器具者以违章论处。各单位要把电力安全工器具的正确使用与管理作为日常工作内容，开工前、收工后、作业中都要把好关。

公司安监部负责制定并及时修订公司电力安全工器具管理办法，编制电力安全工器具购置计划，负责电力安全工器具的选型、选厂购置、验收、试验、使用、保管和报废工作，按规程规定对各单位电力安全工器具进行全面检查试验。

三、电力安全工器具的报废

符合下列条件之一者，即予以报废。一是电力安全工器具经试验或检验不符合国家或行业标准。二是超过有效使用期限，不能达到有效防护功能指标。（对于注明使用期限的电力安全工器具，若超出使用期限必须报废）。

报废的电力安全工器具应及时清理，不得与合格的电力安全工器具混放在一起，更不得使用报废的电力安全工器具。

报废的电力安全工器具以及不能达到有效防护功能指标的电力安全工器具，应及时统计上报备案。

报废的电力安全工器具由分厂负责暂时存放于专门的场所，加强管理，防止报废的电力安全工器具再次流入生产现场。

Task Ⅴ Safekeeping and storage of electric power safety tools and instruments

The safekeeping, storage and scrapping of electric power safety tools and instruments must meet the requirements of national and industry standards and product specifications.

Ⅰ. Safekeeping and storage of electric power safety tools and instruments

(Ⅰ) Rubber and plastic electric power safety tools and instruments

(1) Rubber and plastic electric power safety tools and instruments shall be stored in a dry, ventilated, light-proof environment, more than 20 cm away from the ground and walls and more than 1 m from the heat source, to avoid direct sunshine, or other light sources, to avoid rain and snow, to prevent extrusion, folding and sharp objects from colliding. It is forbidden to store them with oil, acid, alkali or other corrosive materials.

(2) Goggles shall be kept in a clean place that is not easy to collide.

(3) Gas masks shall be stored in a dry, ventilated warehouse free of acid, alkali, solvent and other substances. Heavy pressure is strictly prohibited. The storage period of the canister (box) of the gas mask is 5 years (3 years). Expired products shall not be used until they have passed the inspection.

(4) The air respirator shall be packed in a packing box during storage to avoid prolonged exposure to the sun and cannot be stored together with oleic acid, alkali or other harmful substances. Heavy pressure is strictly prohibited.

(5) Arc-proof clothing must be washed and dried before storage, and shall not be put together with corrosive materials, shall be stored in a dry and ventilated place to avoid long-term contact with the ground moisture, to prevent ultraviolet radiation for a long time. When it is stored for a long time, attention shall be paid to regular drying to avoid mold, insect infestation, and the breeding of bacteria.

(6) Acid-proof clothing made of rubber and plastic shall avoid exposure to high temperature when stored. After use, it shall be cleaned and dried to avoid exposure to sunlight. For long-term preservation, it shall be sprinkled with talcum powder to prevent adhesion. Synthetic fiber acid-proof clothing should not be washed and ironed with hot water to avoid contact with open fire.

(7) Insulating gloves shall be wiped clean and air dried after use, and shall be kept dry and clean. It is best to sprinkle talcum powder to prevent adhesion. Insulating gloves shall be stored in dry, cool cabinets and placed separately from other tools in pairs, as shown in Fig. 1-27. Do not pile anything on the insulating gloves, so as not to pierce the gloves. Insulating gloves are not allowed to be placed in areas that are too cold, overheated, exposed to direct sunlight, or area that contain acids, alkalis, or drugs to prevent gum aging and reduced insulation performance.

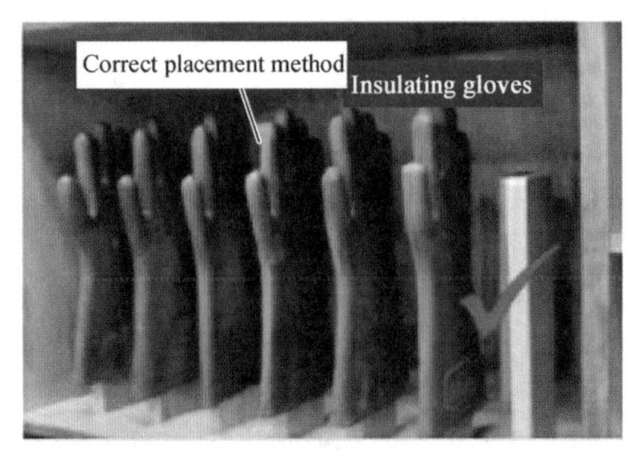

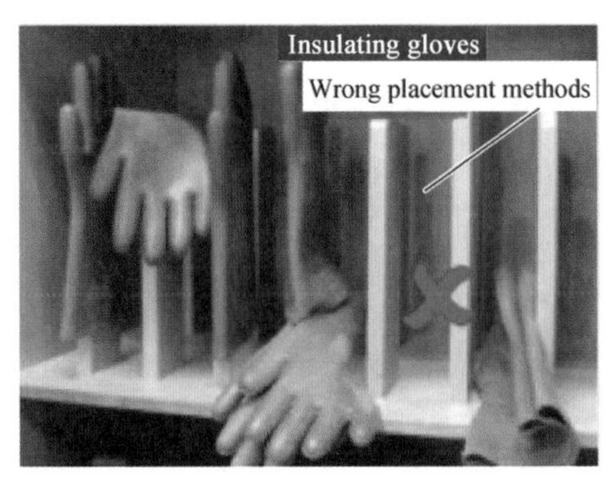

Fig. 1-27 Storage of insulating gloves

(8) Insulating boots (shoes) shall be wiped clean after use, and are not allowed to be placed in areas that are too cold, overheated, exposed to direct sunlight, or area that contain acids, alkalis, oils, and chemicals. They shall be stored in a dry and cool dedicated cabinet or support, placed in pairs and fixed, as shown in Fig. 1-28.

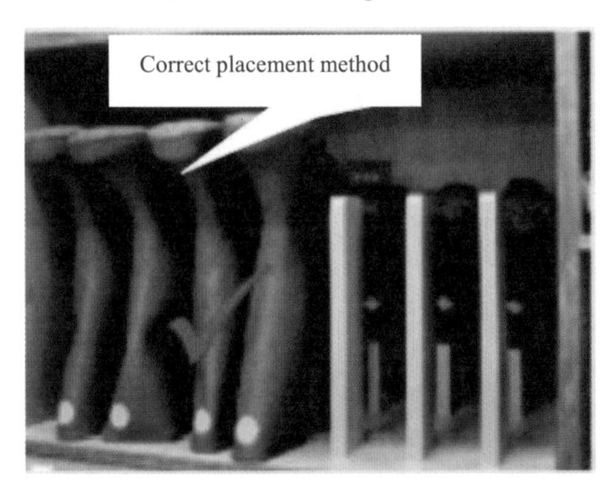

Fig. 1-28 Storage of insulating boots

(9) Acid-proof boots should be rinsed with water immediately, be stored in a cool place, sprinkled with talcum powder to prevent adhesion and should be avoided contact with oils, organic solvents, and sharp objects.

(10) Anti-static shoes and conductive shoes shall be kept clean. If the surface is contaminated with dust, oil wax, insulation material, or an insulation layer is formed due to aging, it will greatly affect its resistance. When brushing, use a soft bristle brush, a soft cloth dipped in alcohol or a neutral detergent that does not contain acid or alkali for wiping.

(11) The insulating protective cover shall be wiped clean after use, be put into the packing bag and placed in a clean, dry and ventilated shelf or special cabinet, on which no objects shall be piled.

(Ⅱ) Epoxy resin electric power safety tools and instruments

Epoxy resin electric power safety tools and instruments shall be stored in a well ventilated, clean and dry place, avoiding direct sunlight, without corrosion and harmful substances, as shown in Fig. 1-29.

Fig. 1-29　Storage of epoxy resin electric power safety tools and instruments

(1) The insulating rod must be cleaned after use, and it shall be mounted on the support or suspended during storage. It shall not directly touch the ground and walls to prevent moisture and dirt, and it shall be placed in a complete set.

(2) Insulating barriers shall be uniformly numbered and stored indoors on a dedicated tool rack or cabinet that is dry, ventilated, and over 20 cm above the ground. If there are slight scratches on the surface, insulation paint shall be applied for treatment.

(3) Reel the soft copper wire when the grounding wire is not in use. Store in a dry room and preferably on a dedicated rack. The number on the rack shall be consistent with the number of the grounding wire.

(4) The phasing tester shall be stored on a dry and ventilated holder or in a dedicated packing box.

(5) After use, electroscope shall be stored in a moisture-proof box or storage cabinet for insulation safety tools and instruments in a ventilated and dry place, as shown in Fig. 1-30.

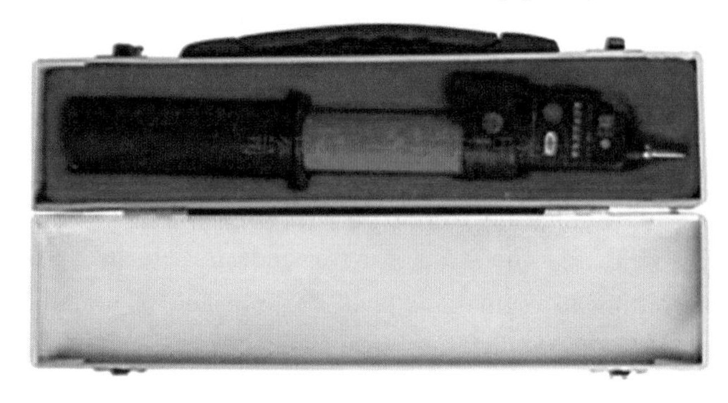

Fig. 1-30　Storage of electroscope

(6) Insulating clamps shall be stored in dedicated boxes or cabinets to isolated moisture and wear.

(Ⅲ) Fiber electric power safety tools and instruments

Fiber electric power safety tools and instruments shall be placed in a dry, ventilated place that avoids direct sunlight, which is free of corrosion and harmful substances, and is kept at a distance of more than 1m from the heat source.

(1) When the safety belt is not in use, it shall be kept by a specially-assigned person. During storage, it shall not come into contact with high temperature, open fire, strong acid, strong alkali or sharp objects, and shall not be stored in a wet place. During storage, the appearance of the safety belt shall be inspected periodically, the abnormal safety belt shall be replaced immediately, and the inspection frequency shall be determined based on the frequency of use of the safety belt.

(2) The safety belt shall be inspected and cleaned regularly after each use.

Ⅱ. Management of electric power safety tools and instruments

Electric power safety tools and instruments shall be stored in a dedicated room of safety tools and instruments, a management ledger for electric power safety tools and instruments shall be established, and a copy shall be submitted to the Company’s Safety Supervision Department. The accounts, cards, and objects shall be consistent, and the test reports, test certificates, and inspection records shall be complete.

Electric power safety tools and instruments shall be kept by dedicated personnel and recorded for requisition. The keeper shall regularly carry out daily inspection, maintenance and service, and those found to be unqualified or beyond the test cycle shall be stored separately, marked with signs reading not allowed to be used, stopped using them, and send them to the Company's Safety Supervision Department for unified scrapping in time. The persons in charge and part-time (full-time) safety officers shall organize inspections of electric power safety tools and instruments before use in accordance with regulations, and keep inspection records. All electric power safety tools and instruments have specific purposes and scope of use, and must be used correctly and properly stored. Misappropriation for other purposes is strictly prohibited.

It is necessary to carry out safety training for the operators, who shall strictly implement the operation regulations, and correctly use the electric power safety tools and instruments. Anyone who is not familiar with the methods of operation shall not use electric power safety tools and instruments.

All organizations shall strictly supervise and manage the correct use of electric power safety tools and instruments, and those who do not wear and use electric power safety tools and instruments correctly according to regulations shall be punished in violation of regulations. All organizations shall take the correct use and management of electric power safety tools and instruments as the content of their daily work, and strictly check before, after and during the operation.

The Company's Safety Supervision Department is responsible for formulating and revising the management measures of the Company's electric power safety tools and instruments in time, for compiling the purchase plan of electric power safety tools and instruments, for the type selection, plant selection and purchase, acceptance, test, use, storage and scrapping of electric power safety tools and instruments, and for conducting a comprehensive inspection and test of the electric power safety tools and instruments of each organization in accordance with the regulations.

Ⅲ. Scrapping of electric power safety tools and instruments

Those who meet one of the following conditions shall be scrapped. First, the electric power safety tools and instruments are not in line with the national or industry standards according to the test or inspection results. Second, the electric power safety tools and instruments exceed the effective service life and fail to meet the effective protective function indicators. (The electric power safety tools and instruments marked with the service life must be scrapped if they exceed the service life.)

Scrapped electric power safety tools and instruments shall be cleaned up in time and shall not be mixed with qualified electric power safety tools and instruments, let alone use scrapped electric power safety tools and instruments.

Scrapped electric power safety tools and instruments and electric power safety tools and instruments that cannot reach the effective protective function indicators shall be counted and reported in time for the record.

Scrapped electric power safety tools and instruments shall be temporarily stored in specialized places by the branch factory, and management shall be strengthened to prevent the scrapped electric power safety tools and instruments from flowing into the production site again.

任务六　电力安全工器具检查、使用及保管实训

一、作业任务

对常用的电力安全工器具——安全帽、接地线、验电器、绝缘手套、绝缘靴、绝缘杆进行检查，并同时口述检查过程。口述内容包括检查、使用及保管项目、检查（检测）结果。

二、引用标准及文件

《国家电网公司电力安全工作规程（变电部分）》，以下简称《安规》。

三、作业条件

作业人员精神状态良好，熟悉工作中安全措施、技术措施以及现场工作危险点。

四、作业前准备

1. 作业前工器具准备

（1）安全帽1顶、接地线1副、验电器1副、绝缘手套1双、绝缘靴1双、绝缘杆一副。工器具、设备等定置摆放。

（2）按《安规》要求正确使用劳动防护用品，并穿戴规范。

2. 危险点及预防措施

危险点：电力安全工器具检查过程中，工器具伤人。

预控措施：工作现场设置安全围栏，进入实训场地佩戴安全帽。

五、作业规范及要求

（1）给定条件：绝缘手套、绝缘靴、××kV高压验电器、安全帽、××kV接地线、绝缘杆。

（2）一人操作，在20 min内完成。

（3）口述检查、使用及维护项目、检查（检测）结果。

（4）操作完毕将工器具摆放整齐。

六、作业流程及标准（见表1-2）

表1-2 《电力安全工器具的检查、使用及保管》评分标准

<table>
<tr><td>班级</td><td></td><td>姓名</td><td></td><td>学号</td><td></td><td>考评员</td><td colspan="2"></td><td>成绩</td><td colspan="2"></td></tr>
<tr><td>序号</td><td>作业名称</td><td colspan="4">质量标准</td><td>分值/分</td><td colspan="3">扣分标准</td><td>扣分</td><td>得分</td></tr>
<tr><td>0</td><td>工作前准备</td><td colspan="4">着装：按《安规》要求正确使用劳动防护用品，并穿戴规范</td><td>1</td><td colspan="3">未按要求着装扣1分</td><td></td><td></td></tr>
</table>

续表

序号	作业名称	质量标准	分值/分	扣分标准	扣分	得分
0	工作前准备	准备好本项目所需安全工器具	1	少准备一项安全工器具扣0.5分，扣完为止		
		报告考官：××考生准备完毕，请求考试	0.5	未报告考官，扣0.5分		
1	绝缘手套					
1.1	检查	检查标签、合格证是否完善，是否在试验合格的有效期内	1	未检查标签、合格证各扣0.5分		
		外观情况：表面有无损伤以及是否清洁，有灰尘和污垢的应擦拭干净，表面损伤的及有烧灼痕迹的不得使用，不得有裂缝、成洞、毛刺、划痕等缺陷	1	未检查扣1分		
		充气试验：将手套朝手指方向卷曲，观察有无漏气或裂口	2	未正确做充气试验扣2分		
		试验周期为6个月（工频耐压试验）	1	回答不正确扣1分		
1.2	使用	应用于1 000 V以上设备时为辅助安全用具，应用于1 000 V以下设备时为基本安全用具	2	回答不正确扣2分		
		戴手套时应将外衣袖口放入绝缘手套的伸长部分	2	未将外衣袖口放入绝缘手套的伸长部分扣2分		
		绝缘手套不能作为一般手套使用	2	回答不正确扣2分		
1.3	保管	使用后必须擦干净，放置处不得直接接触地面、墙面，防止受潮、脏污	2	未清洁扣1分，放置不当扣1分		
		要和其他工器具分开放置，以免损伤绝缘手套	1	放置不当扣1分		
		绝缘手套应成双，定置摆放	1	放置不当扣1分		
2	绝缘靴					
2.1	检查	检查标签、合格证是否完善，并在试验合格的有效期内	1	未检查标签、合格证各扣0.5分		

续表

序号	作业名称	质量标准	分值/分	扣分标准	扣分	得分
2.1	检查	检查外观情况：表面有无损伤及是否清洁，有灰尘和污垢的应擦拭干净，表面损伤的和有烧灼痕迹的不得使用，不得有裂缝、破洞、毛刺、划痕等缺陷	1	未检查扣1分		
		绝缘鞋的使用期限。制造厂规定以大底磨光为止，即当大底露出黄色面胶（绝缘层）就不适合在电气作业中使用了	1	未检查扣1分		
		试验周期为6个月（工频耐压试验）	1	回答不正确扣1分		
2.2	使用	是在任何电压等级的电气设备上工作时，用来使工作人员与地面保持绝缘的辅助安全用具，也是防护跨步电压的基本安全用具	2	回答不正确扣2分		
		使用时应将裤管套入靴筒内；裤管不得长及地面	2	未将裤管套入靴筒内，裤管长及地面扣2分		
		应保持鞋帮干燥	1	回答不正确扣1分		
2.3	保管	使用后必须擦干净，放置处不得直接接触地面、墙面，防止受潮、脏污	2	未清洁扣1分，放置不当扣1分		
		要和其他工器具分开放置，以免损伤绝缘靴	1	放置不当扣1分		
		绝缘靴应成双，定置摆放	1	放置不当扣1分		
3	绝缘杆					
3.1	检查	电压等级应与电气设备或线路的电压等级相符（等于或高于电气设备或线路的电压等级）	1	未检查扣1分		
		检查标签、合格证是否完善，以及是否在试验合格的有效期内	1	未检查标签、合格证各扣0.5分		

续表

序号	作业名称	质量标准	分值/分	扣分标准	扣分	得分
3.1	检查	外观无明显缺陷，绝缘部分和握手部分之间有护环隔开	2	未检查外观，未检查有无护环各扣1分		
		试验周期为12个月（工频耐压试验）	1	回答不正确扣1分		
3.2	使用	要戴绝缘手套，穿绝缘靴	2	未戴绝缘手套，未穿绝缘靴各扣1分		
		应手拿绝缘棒的握手部分，手不可超出护环	2	操作错误扣2分		
		雨雪天操作室外高压设备，应加装防雨罩	2	回答不正确扣2分		
3.3	保管	使用完毕，应擦拭干净后放置，放置处不得直接接触地面、墙面，防止受潮、脏污	2	未清洁扣1分，放置不当扣1分		
		金属制成的工作部分不得触及地面	1	放置不当扣1分		
		要和其他工器具分开放置，以免损伤	1	放置不当扣1分		
		绝缘棒应成套，定置摆放	1	放置不当扣1分		
4	高压验电器					
4.1	检查	额定电压和被测试设备电压等级相一致	1	未检查扣1分		
		分别检查工作触头和绝缘棒的标签、合格证是否完善，并在试验合格的有效期内	1	未分别检查工作触头和绝缘棒的标签、合格证各扣0.5分		
		按压工作触头，初步检查验电器合格	1	未初步检查工作触头扣1分		
		验电前应先将验电器在带电的设备上验电，证实验电器良好	1	回答不正确扣1分		
		绝缘棒外观无明显缺陷，绝缘部分和握手部分之间有护环隔开	1	未检查扣1分		
		工作触头（启动电压试验）和绝缘棒（工频耐压试验）的试验周期为12个月	1	回答不正确扣1分		

续表

序号	作业名称	质量标准	分值/分	扣分标准	扣分	得分
4.2	使用	验电前应先将验电器在带电的设备上验电，证实验电器良好，再对需接地的设备逐相进行验电	4	未先将验电器在带电的设备上验电扣2分，未逐相进行验电扣2分		
		验电器的工作触头不能直接接触带电体，只能逐渐接近带电体，直至验电器发出声、光等报警信号为止	2	将工作触头直接接触带电体扣2分		
		应注意不使验电器受邻近带电体的影响而发出信号	2	使验电器受邻近带电体的影响而发出信号扣2分		
		同杆架设多层电力线路进行验电时，应先验低压，后验高压；先验下层，后验上层；先验距人体较近的导线，后验距人体较远的导线	2	顺序错误扣2分		
		工作人员应戴绝缘手套，穿绝缘靴	2	未戴绝缘手套或未穿绝缘靴均扣2分		
		工作人员应手拿绝缘棒的握手部分，手不可超出护环；人体应与验电设备保持安全距离	2	手超出护环扣1分，与验电设备安全距离不够扣1分		
		雨雪天不得进行室外直接验电	2	回答不正确扣2分		
4.3	保管	使用后必须擦干净，放置处不得直接接触地面、墙面，防止受潮、脏污	1	未清洁扣1分，放置不当扣1分		
		要和其他工器具分开放置，以免损伤	1	放置不当扣1分		
		验电器应成套（工作触头和绝缘棒），定置摆放	1	放置不当扣1分		
5	安全帽					
5.1	检查	检查合格证、生产日期是否完善且在合格期内	1	未检查合格证、生产日期各扣0.5分		
		外观：外壳无龟裂、下凹、裂痕和磨损等情况，发现异常现象要立即更换，不能继续使用。任何受过重击、有龟裂的安全帽，不论有无损坏现象，均不得使用	1	未进行外观检查扣1分		

续表

序号	作业名称	质量标准	分值/分	扣分标准	扣分	得分
5.1	检查	连接部件：帽檐、后箍、透气孔、帽衬接头完好，吸汗带无破损，下颌带、顶衬、顶衬托带无破损、断裂，后箍调节器、下颌带调节器能灵活调节、卡位牢固，顶衬与帽顶的距离在25～50 mm	1	未进行连接部件检查扣1分		
5.2	使用	戴安全帽前应将后箍、下颌带按自己头形调整到适合的位置，然后将帽内弹性带系牢	1	佩戴前未调节后箍、下颌带各扣0.5分		
		双手持帽檐，将安全帽从前至后扣于头顶，调整好后箍，系好下颌带，保证戴好后的安全帽不歪、不晃、不露（不露长发）	2	佩戴不符合要求扣0.5分/项，扣完为止		
		将长头发束好，放入安全帽内（仅适合长发同志）	1	未将长头发放入安全帽内扣1分		
5.3	保管	保持清洁无脏污，置于专用工器具柜或货架，避免受到挤压	1	放置不当扣1分		
		避免存放在酸、碱、高温、日晒、潮湿、有化学溶剂等场所	1	放置不当扣1分		
		严禁与硬物、尖状物放置在一起	1	放置不当扣1分		
6	接地线					
6.1	检查	电压等级：与接地设备电压等级相应，切不可任意取用	1	未检查扣1分		
		检查是否有标签、试验合格证，试验日期应在有效期内，否则不能使用；知晓接地线的试验周期	1	未检查标签、试验合格证各扣0.5分		
		连接部件及外观：软铜线透明护套无严重磨损；铜线无断股、散股、松股，其截面面积不小于25 mm^2；三相合一处连接牢固；螺栓紧固，无松动、滑丝、锈蚀、融化现象；夹具完好无裂纹，弹性正常；绝缘操作棒 表面清洁光滑，无气泡、皱纹、开裂、划伤，绝缘漆无脱落；有护环且护环完好	3	每少检查一项扣0.5分		

续表

序号	作业名称	质量标准	分值/分	扣分标准	扣分	得分
6.2	使用	装接地线之前必须验电，验电位置必须与装设接地线的位置相符	2	未验电或装设位置错误扣2分		
		装设接地线需两人进行，一人操作，一人监护	2	未一人操作一人监护，扣2分		
		操作时保证与相邻带电体足够的安全距离	1	未与相邻带电体保持足够的安全距离扣1分		
		戴绝缘手套，手握在护环以下	1	未戴绝缘手套或手超过护环均扣1分		
		按正确的装拆顺序：装设接地线时，先接接地端，后接导线端；先挂低压，后挂高压；先挂下层，后挂上层。拆接地线时的顺序与此相反	3	未正确装拆接地线，扣3分		
6.3	保管	应保持清洁，存放在干燥的室内专用安全工器具柜内	1	放置不当扣1分		
		每组接地线均应编号，并存放在固定的地点	1	放置不当扣1分		
		存放位置应编号，接地线号码与存放位置号码必须一致	1	放置不当扣1分		
7	工作终结	整理工器具，并将安全用具摆放至初始状态	1	未整理工器具或摆放不整齐扣1分		
		汇报考官，考试完毕	0.5	未汇报考官，扣0.5分		
合计			100			

Task Ⅵ Practical training of inspection, use, and safekeeping of electric power safety tools and instruments

Ⅰ. Operating tasks

Check the commonly used electric power safety tools and instruments: safety helmet, grounding wire, electroscope, insulating gloves, insulating boots, insulating rod, and then verbally describe the inspection, use, and safekeeping items, as well as the inspection (testing) results.

Ⅱ. Referenced standards and documents

Electric Power Safety Working Regulations (Power Transformation) of State Grid Corporation of China. Hereinafter referred to as the *Electric Power Safety Working Regulations*.

Ⅲ. Operating conditions

The operation personnel shall be in a good mental state and familiar with safety measures, technical measures, and dangerous points of field work.

Ⅳ. Preparation before operation

1. Preparation of tools and instruments before operation

(1) One safety helmet, one grounding wire, one electroscope, one pair of insulating gloves, one pair of insulating boots and one insulating rod. Place tools and instruments, equipment, etc. in fixed positions.

(2) Use labour protective equipment correctly according to the requirements of the *Electric Power Safety Working Regulations* and wear them in a standardized manner.

2. Dangerous points and preventive measures

Dangerous points: During the inspection of electric power safety tools and instruments, the tools and instruments hurt people.

Prevention and control measures: Set up safety fences at the work site and wear safety helmets when entering the practical training site.

Ⅴ. Operating specifications and requirements

(1) Given conditions: Insulating gloves, insulating boots, ××kV HV electroscope, safety helmet, ××kV grounding wire, and insulating rod.

(2) One-person operation, completed within 20 minutes.

(3) Verbally describe the inspection, use, and maintenance items, as well as the inspection (testing) results.

(4) Put the tools and instruments neatly after operation.

Ⅵ. Operation processes and standards (see Tab. 1-2)

Tab. 1-2 Scoring standards for *Inspection, Use, and Safekeeping of Electric Power Safety Tools and Instruments*

Class		Name	Student ID	Examiner	Score	
S/N	Operation name	Quality standard	Points	Deduction criteria	Deduction	Score
0	Preparations before work	Dress: Use labour protective equipment correctly according to the requirements of the *Electric Power Safety Working Regulations* and wear them in a standardized manner	1	Deduct 1 point for not dressing as required		
		Prepare the safety tools and instruments required for this item	1	Deduct 0.5 points for missing one safety tool and instrument in terms of preparation, until all points are deducted		
		Report to the examiner: Examinee ×× is ready and requests starting examination.	0.5	Deduct 0.5 points if failing to report to the examiner		
1	Insulating gloves					
1.1	Inspections	Check whether the label and certificate are perfect, and within the period of validity of the test	1	Deduct 0.5 points if failing to check label and certificate, respectively		
		Appearance: Whether the surface is damaged and clean; Those with dust and dirt shall be wiped clean; Those with surface damage and burning marks shall not be used; There must be no cracks, holes, burrs, scratches or other defects	1	Deduct 1 point if failing to check		
		Air inflation test: Curl the gloves toward the fingers to see if there are any air leaks or cracks	2	Deduct 2 points if failing to conduct air inflation test correctly		
		The test cycle is 6 months (power-frequency voltage-withstand test)	1	Deduct 1 point if failing to answer correctly		
1.2	Use	When applied to equipment above 1000 V, it is an auxiliary safety appliance, and when applied to equipment below 1000 V, it is a basic safety appliance	2	Deduct 2 points if failing to answer correctly		
		When wearing gloves, the coat cuffs shall be tucked into the extended part of the insulating gloves	2	Deduct 2 points if failing to tuck the coat cuffs into the extended part of the insulating gloves		
		Insulating gloves cannot be used as general gloves	2	Deduct 2 points if failing to answer correctly		
1.3	Safekeeping	After use, it must be wiped clean and placed in a place that does not directly contact the ground or walls to prevent moisture and dirt	2	Deduct 1 point if failing to cleaning, 1 point for improper placement		
		Separate it from other tools and instruments so as not to damage the insulating gloves	1	Deduct 1 point for improper placement		
		Insulating gloves shall be placed in pairs	1	Deduct 1 point for improper placement		

Continued

S/N	Operation name	Quality standard	Points	Deduction criteria	Deduction	Score
2	Insulating boots					
2.1	Inspections	Check whether the label and certificate are perfect, and within the period of validity of the test	1	Deduct 0.5 points if failing to check label and certificate, respectively		
		Check the appearance: Whether the surface is damaged or clean. Those with dust and dirt shall be wiped clean. Those with surface damage and burning marks shall not be used, and there shall be no defects such as cracks, holes, burrs, and scratches	1	Deduct 1 point if failing to check		
		With regard to the service life of insulating shoes, the manufacturer stipulates that the service life of insulating shoes ends until the bottom is polished, which means that when the yellow surface adhesive (insulation layer) is exposed on the bottom, it is no longer suitable for use in electrical operation	1	Deduct 1 point if failing to check		
		The test cycle is 6 months (power-frequency voltage-withstand test)	1	Deduct 1 point if failing to answer correctly		
2.2	Use	It is an auxiliary safety appliance used to keep personnel insulated from the ground when working on electrical equipment of any voltage class, and also a basic safety appliance for protecting from step voltage	2	Deduct 2 points if failing to answer correctly		
		The trouser legs shall be tucked into the boots when in use; The trouser legs shall not reach the ground	2	Failing to tuck the trouser legs into the boots; Deduct 2 points if the trouser legs reach the ground		
		Keep the upper dry	1	Deduct 1 point if failing to answer correctly		
2.3	Safekeeping	After use, it must be wiped clean and placed in a place that does not directly contact the ground or walls to prevent moisture and dirt	2	Deduct 1 point if failing to cleaning, 1 point for improper placement		
		Separate it from other tools and instruments so as not to damage the insulating boots	1	Deduct 1 point for improper placement		
		Insulating boots shall be placed in pairs	1	Deduct 1 point for improper placement		
3	Insulating rod					
3.1	Inspections	The voltage class shall be in line with the voltage class of the electrical equipment or line (equal to or higher than the voltage class of the electrical equipment or line)	1	Deduct 1 point if failing to check		
		Check whether the label and certificate are perfect, and within the period of validity of the test	1	Deduct 0.5 points if failing to check label and certificate, respectively		
		There is no obvious defect in appearance, and there is a protective ring between the insulating part and the handle part.	2	Deduct 1 point if failing to inspect appearance and presence of protective ring		

Continued

S/N	Operation name	Quality standard	Points	Deduction criteria	Deduction	Score
3.1	Inspections	The test cycle is 12 months (power-frequency voltage-withstand test)	1	Deduct 1 point if failing to answer correctly		
3.2	Use	Wear insulating gloves and insulating boots	2	Deduct 1 point if failing to wear insulating gloves and insulating boots, respectively		
		The handle part of the insulating rod shall be held in hand, and the hand shall not exceed the protective ring	2	Deduct 2 points for wrong operation		
		For operating outdoor HV equipment on rainy and snowy days, rain covers shall be installed	2	Deduct 2 points if failing to answer correctly		
3.3	Safekeeping	After use, it should be wiped clean and placed in a place that does not directly contact the ground or walls to prevent moisture and dirt	2	Deduct 1 point if failing to cleaning, 1 point for improper placement		
		The working part made of metal shall not touch the ground	1	Deduct 1 point for improper placement		
		Separate it from other tools and instruments to avoid damage	1	Deduct 1 point for improper placement		
		Insulating rods shall be placed in pairs	1	Deduct 1 point for improper placement		
4	HV electroscope					
4.1	Inspections	The rated voltage shall be consistent with the voltage class of the equipment under test	1	Deduct 1 point if failing to check		
		Check whether the labels and certificates of the working contact terminal and insulating rod are complete and within the period of validity of the test	1	Deduct 0.5 points if failing to check the labels and certificates of the working contact terminal and insulating rod, respectively		
		Initially check if the electroscope is qualified by pressing the working contact terminal	1	Deduct 1 point if failing to initially check the working contact terminal		
		Before the electricity detection, the electroscope shall be tested on the live equipment to ensure that the electroscope is in good condition	1	Deduct 1 point if failing to answer correctly		
		There is no obvious defect in the appearance of the insulating rod, and there is a protective ring between the insulating part and the handle part	1	Deduct 1 point if failing to check		
		The test cycle of working contact terminal (starting voltage test) and insulating rod (power-frequency voltage-withstand test) is 12 months	1	Deduct 1 point if failing to answer correctly		
4.2	Use	Before the electricity detection, the electroscope shall be tested on the live equipment to ensure that the electroscope is in good condition, then conduct phase by phase electricity detection on the equipment that needs to be grounded	4	Deduct 2 points if failing to test the electroscope on the live equipment first, and deduct 2 points if failing to conduct electricity detection phase by phase.		

Continued

S/N	Operation name	Quality standard	Points	Deduction criteria	Deduction	Score
4.2	Use	The working contact terminal of the electroscope can not directly contact the charged body, but can only gradually approach the charged body until the electroscope emits sound and light alarm signals	2	Deduct 2 points if the working contact terminal directly contacts the charged body		
		Attention shall be paid to prevent electroscope from sending out signals due to the influence of adjacent charged bodies	2	Deduct 2 points for causing the electroscope sending out signals due to the influence of adjacent charged bodies		
		For the electricity detection of multi-layer electric power lines erected on the same pole, LV shall be tested first and HV later. the lower layer shall be tested first and upper layer later; The conductor closer to the human body shall be tested first, and the conductor far away from the human body later	2	Deduct 2 points if sequence is not correct		
		Workers shall wear insulating gloves and insulating boots	2	Deduct 2 points if failing to wear insulating gloves and insulating boots		
		The worker shall hold the handle part of the insulating rod in hand, and the hand shall not exceed the protective ring. The human body shall maintain a safe distance from electricity detection equipment	2	Deduct 1 point if the hand exceeds the protective ring, and deduct 1 point if the safe distance from the electricity detection equipment is not enough		
		Outdoor direct electricity detection is not allowed on rainy and snowy days	2	Deduct 2 points if failing to answer correctly		
4.3	Safekeeping	After use, it must be wiped clean and placed in a place that does not directly contact the ground or walls to prevent moisture and dirt	1	Deduct 1 point if failing to cleaning, 1 point for improper placement		
		Separate it from other tools and instruments to avoid damage	1	Deduct 1 point for improper placement		
		Electroscopes shall be placed in pairs (working contact terminals and insulating rods)	1	Deduct 1 point for improper placement		
5	Safety helmet					
5.1	Inspections	Check if the certificate and production date are complete and within the qualified period	1	Deduct 0.5 points for failing to check label and production date, respectively		
		Appearance: The housing is free of chapping, depressions, cracks, and wear. If any abnormalities are found, it shall be replaced immediately and cannot be used again. Any hard-hit, cracked safety helmet, regardless of whether it is damaged or not, shall not be used	1	Deduct 1 point if failing to check appearance		

Continued

S/N	Operation name	Quality standard	Points	Deduction criteria	Deduction	Score
5.1	Inspections	Connecting components: The brim, rear hoop, and vent hole of the safety helmet. The cap liner joint is intact, the sweat absorbing belt is not damaged, the jaw band, the top lining and the top lining bracket are not damaged or broken, the rear hoop regulator and the jaw band regulator can be adjusted flexibly, the clamping position is firm, and the distance between the top liner and the top shall be between 25-50 mm	1	Deduct 1 point if failing to check connecting component		
5.2	Use	Before wearing a safety helmet, the rear hoop and the jaw band shall be adjusted to a suitable position according to own head shape, and then the elastic belt inside the helmet shall be fastened	1	Deduct 0.5 points if failing to adjust the rear hoop and the jaw band before wearing.		
		Hold the brim of the helmet in both hands, buckle the safety helmet from front to back on the top of head, adjust the rear hoop, and fasten the jaw belt. Ensure that the safety helmet after wearing is not crooked, swayed, or exposed (without revealing long hair)	2	Deduct 0. 5 points per item for wearing that do not meet the requirements, until all points are deducted		
		Tie up long hair and place it in the safety helmet (only suitable for long haired workers)	1	Deduct 1 point if failing to put long hair into the safety helmet		
5.3	Safekeeping	Keep clean and free from dirt, and place it in a special tools and instruments cabinet or shelf to avoid being squeezed	1	Deduct 1 point for improper placement		
		Avoid storing in places with acid, alkali, high temperature, sunlight, humidity, chemical solvents, etc.	1	Deduct 1 point for improper placement		
		It is strictly forbidden to put it together with hard objects or sharp objects	1	Deduct 1 point for improper placement		
6	Grounding wire					
6.1	Inspections	Voltage class: Corresponding to the voltage class of grounding equipment, it cannot be arbitrarily used	1	Deduct 1 point if failing to check		
		Check if there are labels and test certificates, and the test date shall be within the validity period, otherwise it cannot be used; Know the test cycle of the grounding wire	1	Deduct 0.5 points if failing to check label and test certificate, respectively		

Continued

S/N	Operation name	Quality standard	Points	Deduction criteria	Deduction	Score
6.1	Inspections	Connecting component and appearance: The transparent sheath of the soft copper wire is not severely worn. The copper wire shall have no broken, separate, or loose strands, and its cross sectional area shall not be less than 25 mm^2. The three-phase integration point is firmly connected. The bolts are tightened without looseness, slipping, rust, or melting. The grip is intact without cracks and has normal elasticity; Insulating rod surface shall be clean and smooth, with no bubbles, wrinkles, cracks, and scratches, and the insulation paint shall not fall off. There is a protective ring and the protective ring is intact	3	Deduct 0.5 points for each missed inspection item		
6.2	Use	Electricity detection must be done before the grounding wire is installed, and the position of the electricity detection must be consistent with the position of the grounding wire installed	2	Deduct 2 points if failing to conduct electricity detection or for wrong installation position		
		Installing the grounding wire two persons, one to operate and one to monitor	2	Deduct 2 points if failing to make one person operate and one person monitor		
		Ensure a sufficient safe distance from the adjacent charged body during operation	1	Deduct 1 point if failing to ensure a sufficient safe distance from the adjacent charged body		
		Wear insulating gloves and hold the hand below the protective ring	1	Deduct 1 point if failing to wear insulating gloves or hand exceeding the protective ring		
		Follow the correct installation and disassembly sequence: When installing the grounding wire, first connect the grounding terminal, and then connect the conductor terminal; first LV, then HV; first lower layer, then upper layer. The sequence to disassemble a grounding wire is the opposite	3	Deduct 3 points if the grounding wire is not installed or disassembled correctly		
6.3	Safekeeping	It shall be kept clean and stored in a dry indoor cabinet for special safety tools and instruments	1	Deduct 1 point for improper placement		
		Each group of grounding wires shall be numbered and kept in a fixed place	1	Deduct 1 point for improper placement		
		The storage positions shall be numbered, and the grounding wire number shall be the same as the storage location number	1	Deduct 1 point for improper placement		
7	End of work	Tidy up tools and instruments, and place safety appliances in their initial state	1	Deduct 1 point if failing to tidy up tools and instruments or placing them untidily		
		Report to the examiner that the examination is over	0.5	Deduct 0.5 points if failing to report to the examiner		
Total			100			

模块二 电力测量仪表的检查和使用

电力测量广泛应用于电力系统中，在电气设备的运行和检修过程中，都离不开电力测量。通过电力测，学习者可以了解电气设备的特性和运行情况，由此可见，正确掌握电力仪表与测量的基本知识和技能是十分重要的。

学习目标：

（1）能说出常用电力仪表的名称和测量对象。

（2）能正确使用万用表进行测量。

（3）能正确使用钳形电流表进行测量。

（4）能正确使用绝缘电阻表进行测量。

（5）能正确使用接地电阻表进行测量。

Module Ⅱ Inspection and use of electric measuring instrument

Electric power measurement is widely used in power system. In the operation and maintenance process of electrical equipment, electric power measurement is indispensable. Through electric power measurement, the characteristics and operation of electrical equipment can be understood. It can be seen that it is very important to correctly grasp the basic knowledge and skills of electric power instruments and measurement.

Learning Objectives

(1) Can name commonly used power instruments and their measurement objects.

(2) Be able to use a multimeter correctly for measurement.

(3) Be able to use a clamp ammeter correctly for measurement.

(4) Be able to use an insulation resistance meter correctly for measurement.

(5) Be able to use a grounding resistance meter correctly for measurement.

任务一　万用表的检查和使用

电力测量中常用的万用表，用于测量交流电流、电压，电阻，直流电流等，分为模拟式万用表和电子式数字万用表两种。模拟式万用表，用来指示测量值的是一个动圈式直流电流表，各项测量都转换成动圈式直流电流表的驱动电流，用直流电流表指针的偏转角度来指示被测量的值。此外，还设有分流器（用以扩大电流的测量范围）、倍率器（用以扩大电压的测量范围）、整流器（将交流变成直流）、电池（为测量电阻时提供电源）、切换开关等部分。相对电子式数字万用表来说，模拟式万用表的指针表读取精度较差，但指针摆动过程比较直观，其摆动速度和幅度有时也能比较客观地反映测量值的大小。而电子式数字表万用表读数直观，但数字变化的过程看起来很复杂，不易观看。

图 2-1 所示是一种模拟式万用表的外形图，图 2-2 是它的刻度盘，标示出了各部分的功能。

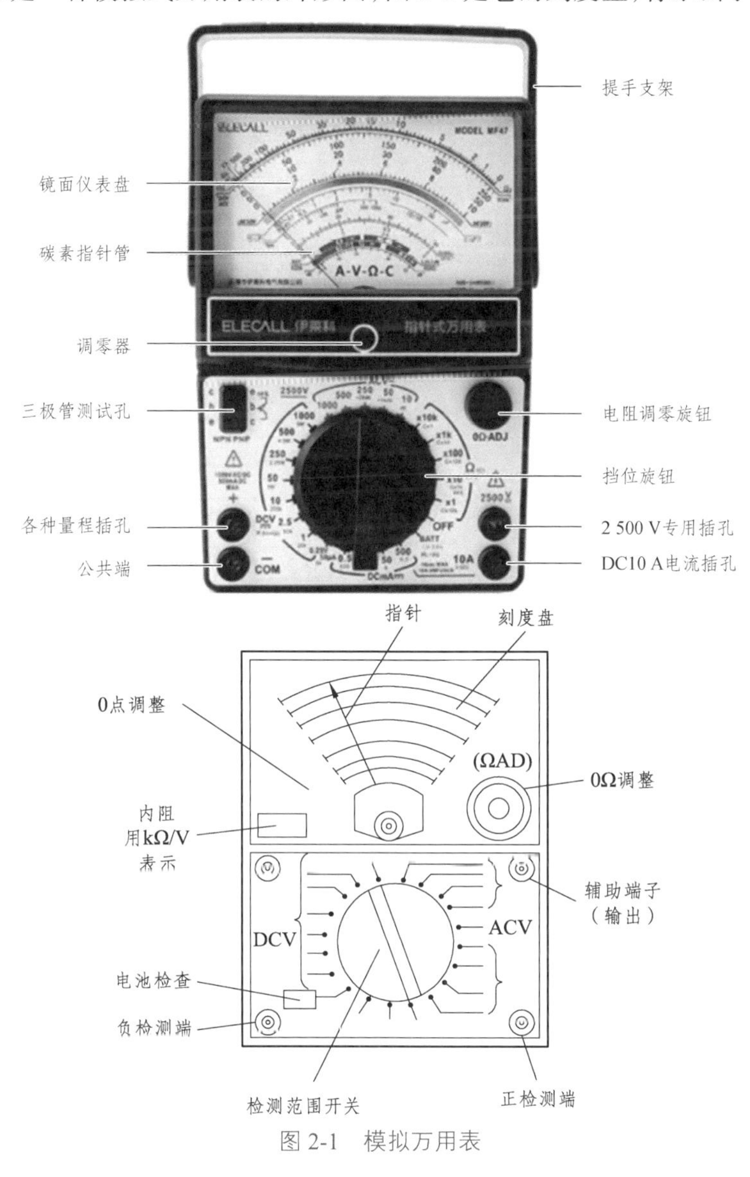

图 2-1　模拟万用表

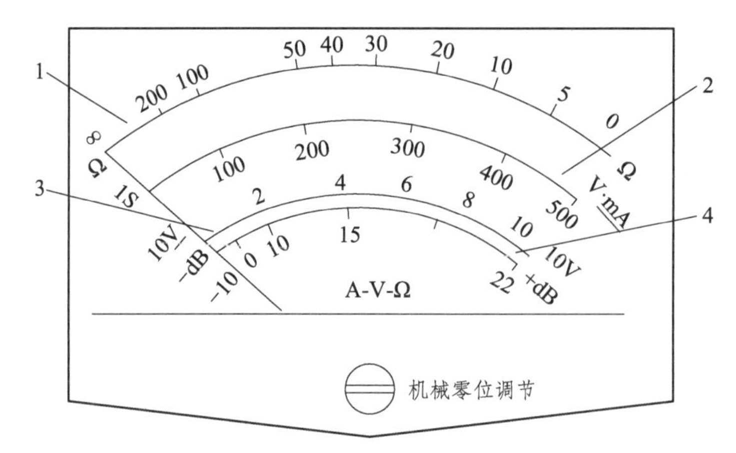

1—电阻 Ω 刻度；2—电压、电流共用刻度；3—10 V 专用刻度；4—dB 数刻度。

图 2-2　模拟万用表刻度盘

一、万用表概述

（一）万用表各主要部分的功能

1. 表头校正器

万用表表笔开路时，表的指针应指在 0 的位置（表盘的左侧，电压和电流刻度的 0 值），如果不在 0 的位置，可用螺丝刀微调使指针处于 0 位，此调整又称零位调整。

2. 测量范围切换开关

测量电流、电压、电阻或是其他不同的测量范围均由这个开关进行转换，各位置的数字以表盘上满刻度的值表示。

3. 零欧姆调整

测量电阻时先将两支表笔短路，这时表针应指向 0（表盘的右侧，电阻刻度的 0 值），如果不在 0 位，可微调此钮（电位器），使表针指向 0。

4. 测量用端子

万用表的附件有两个表笔。测量时其中红色的表笔插到“+”端，黑色的表笔插到符号“−”端。

（二）万用表的性能

万用表的性能如表 2-1、2-2 所示。

表 2-1　万用表的最大刻度值

测量项目	最大刻度值
直流电压/V	0.25、1、2.5、10、50、250、1 000（内阻 20 kQ/V）
交流电压/V	1.5、10、50、250、1 000（内阻 20 kft/V）
直流电流/mA	3 000，30 000，300 000
音频电平/dB	0 ~ +22（AGIOV 范围）

表 2-2　万用表的误差

测量项目	允差值
直流电压、电流	最大刻度值的±3%
交流电压	最大刻度值的±4%
电阻	度盘长度的±3%

二、直流电流的测量

（一）测量直流电流的基本原理

一个磁电式表头就是一个电流表，只不过它的量程为七（一般为几微安至几十微安），若要测较大的电流时，在表头两端并联一个适当阻值的电阻即可，如图 2-3 所示。其中 R_s 称为分流电阻（分流器），阻值的大小可用下式计算：

$$R_s=R_g/(n-1) \tag{2-1}$$

式中 $n=I/I_n$，表示表头量程扩大的倍数。

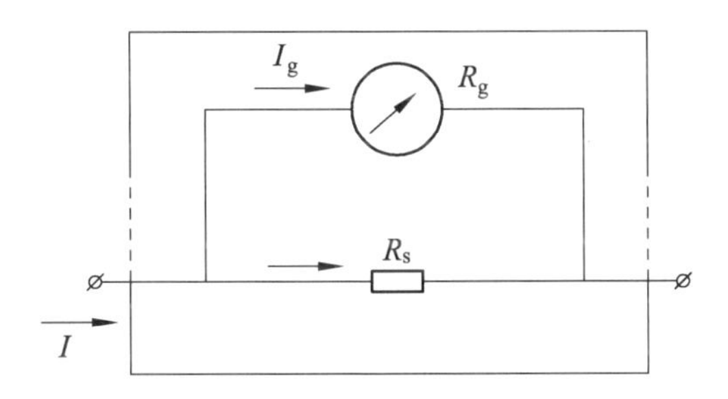

图 2-3　单量程电流表原理图

当 R_s 为定值时，被测电流 I 与流过表头的电流的大小成一定的比例关系，因此表头指针的偏转角可以反映被测电流的大小。

多量程的电流表，即在表头两端并联上不同阻值的电阻，由转换开关接入电路。分流器有两种连接方法：独立分挡和闭路抽头。从保护表头的安全因素出发，若开关接触不良，独立分挡连接式可能损坏表头，而闭路抽头连接式可避免损坏表头，因此电路连接时多采用闭路抽头连接式，如图 2-4 所示。此电流表有三挡量程，分别为 I_1、I_2、I_3。

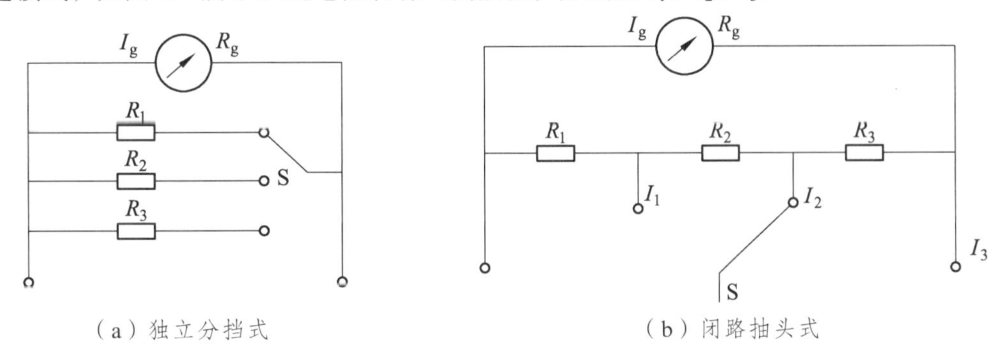

（a）独立分挡式　　（b）闭路抽头式

图 2-4　多量程电流表原理图

（二）测量直流电流的实际操作

测量电路的直流电流则需要切断被测部位的电路后，将万用表串接在电路之中，如图 2-5 所示。

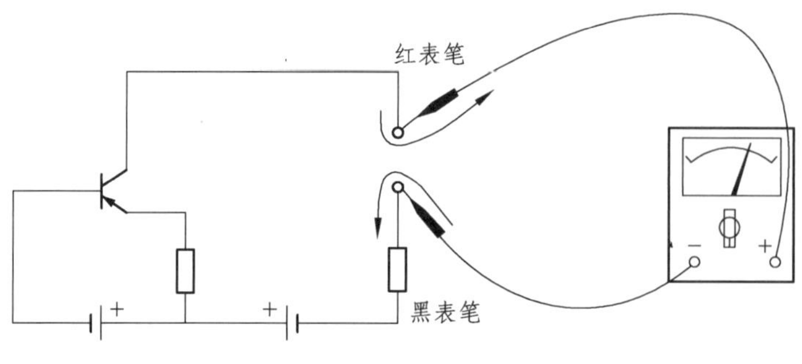

图 2-5 直流电流的测量

测量时，将测量范围钮旋至 DC mA 的位置，接通电路后，根据测量值再进一步选择测量范围。如图 2-5 所示，测量时，将万用表的正端（红色表笔）接到电压高的一端，万用表负端（黑色表笔）接到电压低的一端。如果极性接反，表针会向反方向偏摆，有可能引起万用表故障，这是需要注意的。

三、直流电压的测量

（一）测量直流电压的基本原理

用一个单独的磁电式表头就可测量小于 U_g（$U_g=I_gR_g$）的直流电压。若要测较大的电压，根据串联电阻可以分压的原理，在表头上串联一个适当的电阻即可，如图 2-6 所示。

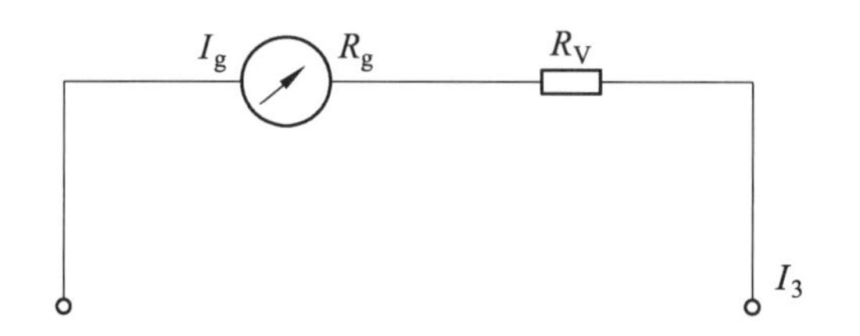

图 2-6 单量程直流电压测量

图中 R_V 为分压电阻，阻值大小用式（2-2）计算：

$$R_V=(U-I_gR_g)/I_g=(mU_g-I_gR_g)/I_g=(m-1)I_g \tag{2-2}$$

式中，$m=U/U_g$ 时表示表头量程扩大的倍数。

当 R_V 为定值时，被测电压 U 与流过表头的电流的大小成一定的比例关系，因此表头指针的偏转角可以反映被测电压的大小。

若多个分压电阻与表头串联，就可制成多量程的直流电压表。连接方式分为单用式和共用式，电路原理如图 2-7 所示。

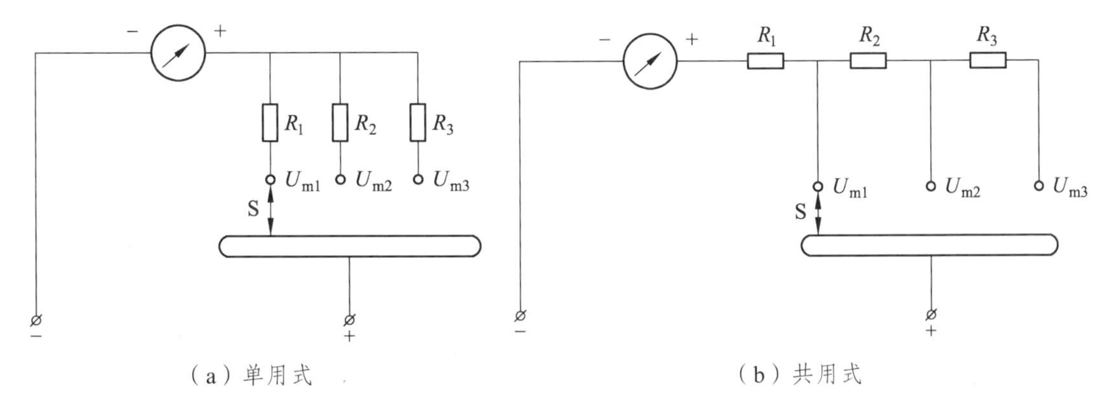

（a）单用式　　　　（b）共用式

图 2-7　多量程直流电压测量

（二）测量直流电压的实际操作

测量电路中的直流电压，要先将测量范围切换开关置于直流电压挡，并选择适当的测量范围。测量实例如图 2-8 所示，需要测量集电极负载电阻上的压降。将万用表的正端（红色表笔）接到电压高的一端，万用表负端（黑色表笔）接到电压低的一端。如果极性接反，表针会向反方向偏摆，有可能引起万用表故障。

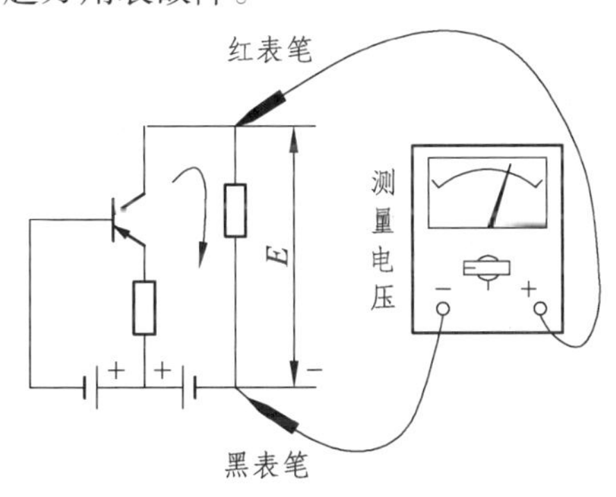

图 2-8　直流电压的测量

四、交流电压的测量

（一）测量交流电压的基本原理

由于万用表的表头是磁电系测量机构，只能测量直流。因此测量交流电时，必须采取整流措施，将交流电转变为直流电。对交流信号进行整流的方式有多种，最常见的是平均整流和峰值整流，而在万用表上普遍采用平均值整流方式，而平均值整流又可分为半波整流和全波整流两种。

（1）半波整流电路：是一种利用二极管的单向导通性进行整流的常见电路，除去半周、剩下半周的整流方法叫半波整流。如图 2-9 所示，整流二极管 VD_1 与表头串联构成一个支路，而二极管 VD_2 并接在由表头和 VD_1 串联的支路两端。由于二极管的单向导通性，输入信号经 VD_1、VD_2 的作用，流过表头的电流是单向脉动电流，其波形如图 2-9（a）（b）所示。当 VD_2 导通时，二级管的正向电阻很低，所以 a、b 端的电压也很低，一般只有 0.3 ~ 0.7 V。这就保护了 VD_1，使它不会被反向电压击穿。

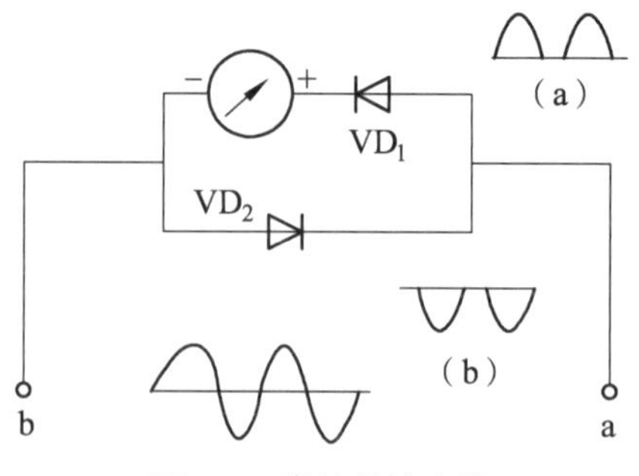

图 2-9　半波整流电路

（2）全波整流电路：如图 2-10 所示，它是由四个整流二极管组成的桥式整流电路，二极管分别成为桥路的四个臂。两个对角线一个接交流电源，另一个接磁电系测量机构。由于二极管 VD_1、VD_3 和 VD_2、VD_4 的作用，在交流电压的一个周期内，表头中流过的是两个同方向的半波电流，其波形如图 2-10（a）所示。外加的交流电压的数值相等时，在全波整流电路中，流过表头的电流要比半波整流电路大一倍。所以全波整流电路比半波整流电路有较高的灵敏度，或者说其整流效率要高一倍。

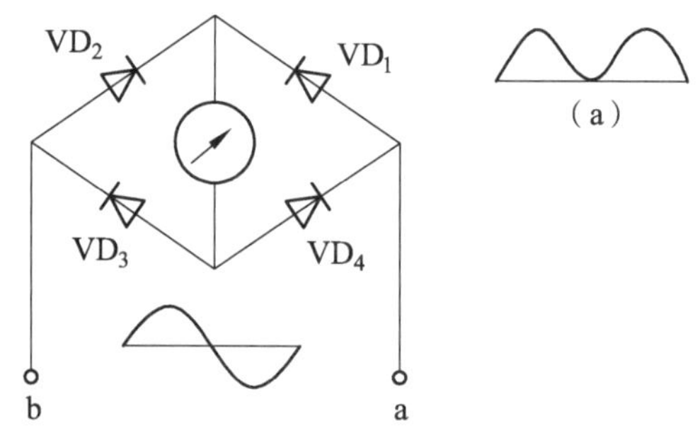

图 2-10　全波整流电路

（3）多量程交流电压表：将带有整流电路的表头电路串联各种数值的附加电阻，即构成多量程交流电压表。与直流电压挡的测量线路类似，多量限交流电压挡的测量线路也分为单用式和共用式两种，如图 2-11 所示。

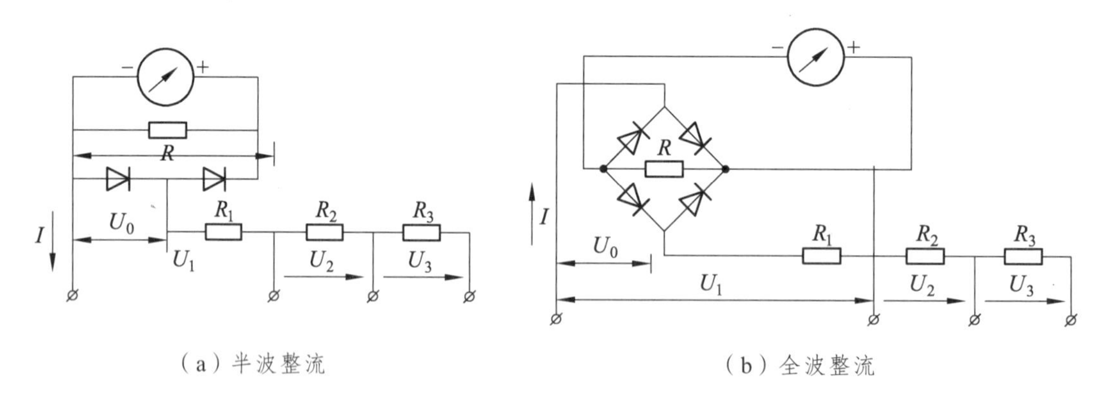

图 2-11　整流式交流电压表电路

（二）测量交流电压的实际操作

电源变压器抽头电压的测量，显示器灯丝电压的检查，交流 220 V 电压的检查，都属于交流电压的测量范围。测量交流电压时，将测量范围开关旋至 AC 挡，再进一步选择测量范围。表笔的极性任意。

如果测量叠加在直流电压上的交流分量可在表笔上串接一只 0.1 MF 的电容，以便隔离直流分量，有些万用表中设有内置电容。一般不能测 50 kHz 以上的交流信号。

五、直流电阻的测量

（一）测量直流电阻的基本原理

万用表的电阻挡基本电路如图 2-12 所示。

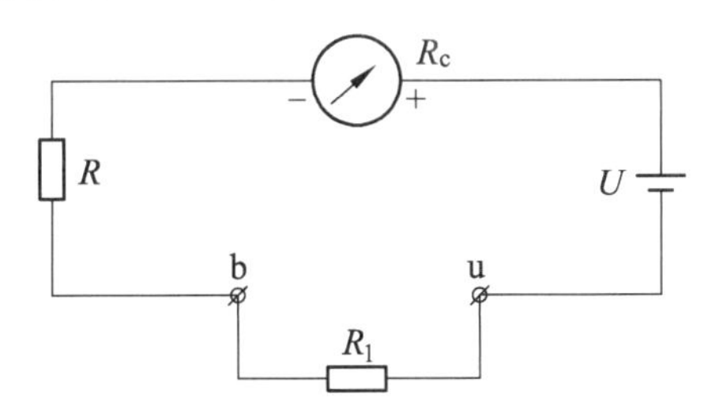

图 2-12　万用表欧姆挡测量电阻的原理电路图

根据欧姆定律可知，流过被测电阻的电流为

$$I=\frac{U}{R_c+R+R_x} \tag{2-3}$$

由上式可知，当电池电压 U、固定电阻 R、R_c 不变时，流过表头电流的大小与被测电阻 R_x 的大小是一一对应的，因此表头指针的偏转角可以用来反映被测电阻的大小。同时可以得出以下几点结论。

（1）根据电阻测量原理可知，流过表头的电流与被测电阻的关系不是线性关系，因此，欧姆表标度尺的刻度是不均匀的。当 R_x 为无穷大时，I=0，指针偏转角为 0；当被测电阻 R_x=0 时，流过表头的电流 I 恰好是表头的满偏电流 I_c，这时指针满刻度偏转。可见欧姆档标度尺为反向刻度，与其电流、电压挡的标度尺的刻度方向恰好相反，如图 2-13 所示。

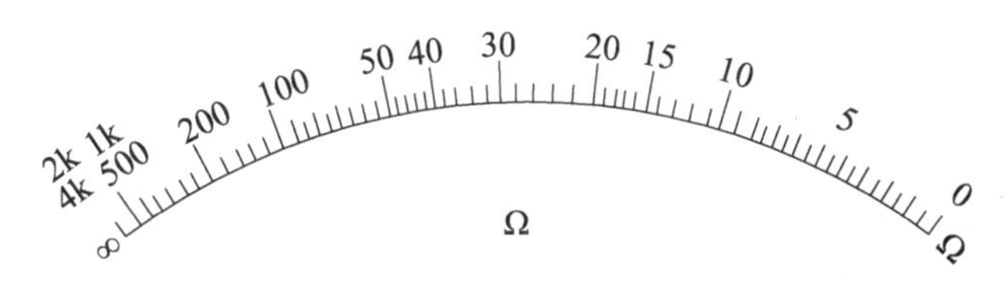

图 2-13　欧姆档标度尺

（2）万用表欧姆挡随着使用或者存放时间的增加，电池端电压会逐渐下降，必然会使测量时工作电流减小，从而造成测量误差。最明显的误差就是当 R_x=0 时，表头指针不能达到满刻度偏转，即不能达到零欧姆刻度线。因此，实际的万用表欧姆挡中都设有零欧姆调整器，常用的零欧姆调整器一般都采用分压式电路。通过调节电位器来调节分流电流的大小，从而确保当 R_x=0 时，流过表头的电流等于表头的满偏电流，使指针达到零欧姆刻度线。

（3）如图 2-12 所示，当 R_x=R_c+R 时，有

$$I=\frac{U}{R_c+R+R_x}=\frac{U}{2(R_c+R)}=\frac{I_c}{2} \tag{2-4}$$

式中，I_c 为表头的满偏电流，当表头指针在标度尺的中心位置时，所指示的数值称欧姆中心值。

它所指示的欧姆数值正好是该量限的总内阻值。首先由于欧姆表标度尺的刻度不均匀，知道了欧姆中心值，就确定了欧姆表的有效测量范围，使得其有效测量范围一般为（$\frac{1}{10}$～10）倍欧姆中心值。如果被测电阻超出该范围太大，则需要改变量程。其次根据欧姆中心值，可以按十进制倍数扩大其量程。这样做可以使各个量程共用一条标度尺，使读数方便。

（4）测量电阻量程的扩大，欧姆表的总内阻增大，测量时表头的工作电流相应减小，当 R_x=0 时，指针不能指到零刻度。为了解决这一问题，第一，在电池电压不变的情况下，改变测量电路的分流电阻，以适应不同量程时对工作电流的要求。第二，提高工作电压，高电阻挡时电表的内阻增加，但提高了电池电压后，当 R_x=0 时，仍可保持表头电流达到满偏，可以通过转换开关接入较高的电池电压。

（二）测量直流电阻的实际操作

测量电阻值时，先将测量范围开关旋至欧姆挡，再根据电阻值选择测量范围。测量电阻时先将两表笔短路使表针指在 0 Ω 处，再进行实际电阻的测量，表笔极性任意。如果测量具有大电容电路中的电阻，电容上的充电电荷必须放掉以后再测量。

六、电子式数字万用表

（一）电子式数字万用表的面板结构

电子式数字万用表的面板结构主要包括液晶显示屏、电源开关、功能量程 h_{FE} 插口和输入插孔。以 DT-830 型电子式数字万用表为例，其面板如图 2-14 所示。

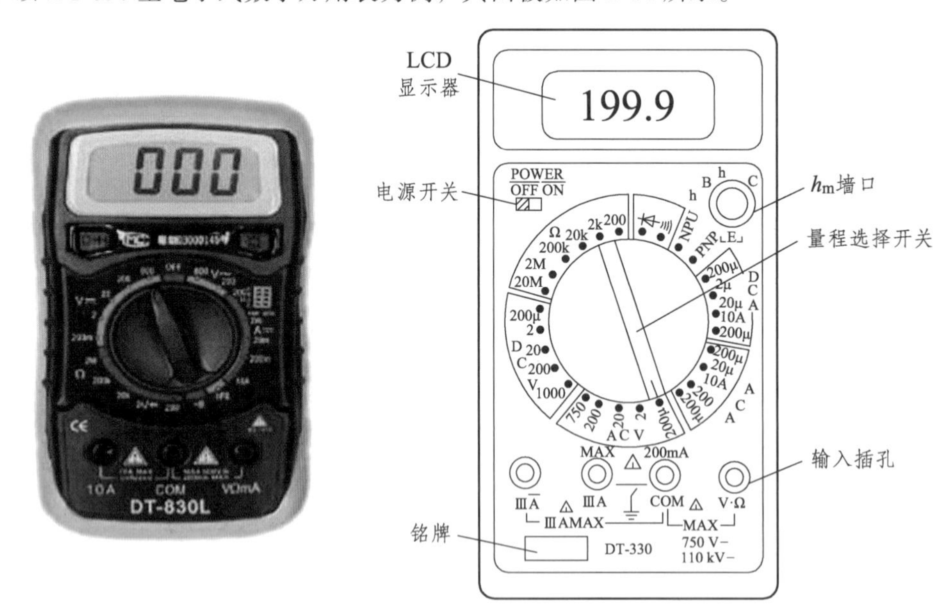

图 2-14　DT-830 型电子式数字万用表

DT-830 型电子式数字万用表面板由五部分组成，各部分的名称和作用如下：

（1）LCD（液晶显示屏）：显示各种被测量的数值，包括小数点、正负号及溢出状态。

（2）电源开关：接通和切断表内电池电源。

（3）功能量程开关：根据具体情况转换不同的测量功能和量程。

（4）h_{FE}插口：用来进行三极管参数的测量。

（5）输入插孔：用来外接测试表笔。

（二）电子式数字万用表的主要功能及技术指标

1. 主要功能

电子式数字万用表除和模拟式万用表一样能进行直流电流、电压的测量，交流电流、电压的测量，电阻的测量外，还能够进行二极管结电压的测量，三极管晶体管的测量和线路通断的测量。

2. 技术指标

（1）测量准确度：电子式数字万用表的误差主要包括基准源误差、输入放大器误差、非线性误差和量化误差。

（2）分辨力：电子式数字万用表末位一个字所对应的数值。

（3）输入阻抗：电子式数字万用表处于工作状态下，从输入端看进去的输入电路的等效阻抗。

（4）测量速度：在单位时间内，按规定的准确度完成测量的次数。

（5）响应时间：输入信号发生突变的瞬间到满足准确度的新的稳定显示值之间的时间间隔。

（三）电子式数字万用表的使用方法

1. 面　板

（1）显示屏采用大字号 LCD 显示屏。仪表具有自动调零和自动显示极性的功能。如果被测电压或者电流的极性为负，就会在显示值前面出现负号“-”。当叠层电池的电压低于 7 V 时，显示屏的左上方显示低电压指示符号，提示需要更换电池。超量程时显示“1”或“-1”，视被测电量的极性而定。小数点由量程开关进行同步控制，使小数点左移或者右移。

（2）电源开关。在 POWER 下边标注有符号“OFF（关）”和“ON（开）”。把电源开关拨至“ON”，接通电源，即可使用仪表；使用完毕后应将开关拨至“OFF”位置，以免空耗电池。

（3）功能量程开关可完成测试功能和量程的选择。

（4）h_{FE}插口采用四芯插座，上面标有 B、C、E。E 孔共有两个，在内部连通。测量晶体三极管如h_{FE}值时，应将三个电极分别插入 B、C、E 孔。

（5）输入插孔共有四个，分别标有“10A”“mA”“COM”和“VΩ”。在“VΩ”与“COM”之间标有“MAX750 V ~”“1 000 V−”的字样，表示从这两个孔输入的交流电压不得超过 750 V，直流电压不得超过 1 000 V。另外在“mA”与“COM”之间标有“MAX200 mA”，在“10 A“与“COM”之间还标有“MAX10 A”字样，分别表示输入的交、直流电流的最大允许值。

2. 直流电压的测量

（1）直流电压有五挡，分别为：200 mV、2V、20 V、200 V、1 000 V。

（2）将电源开关拨至“ON”，量程开关拨至“DCV”范围内的合适挡位。

（3）红色表笔接“V · Ω”插孔，黑色表笔接“COM”插孔，表笔与被测电路并联。

（4）最大允许输入电压：1000 VDC（200 mV、2V、20 V 量程）；1100 VDC（200 V、1000 V 量程）。

3. 交流电压的测量

（1）交流电压有五挡，分别为：200 mV、2V、20 V、200 V、750 V。

（2）将量程开关拨至“AC V”，范围内的合适挡位，表笔接法同直流电压的测量。

（3）要求被测电压频率为 45 ~ 500 Hz，最大允许输入电压为 750 V（有效值）。

4. 直流电流的测量

（1）直流电流有四挡，分别为：200 μA、2 mA、20 mA、200 mA。

（2）将量程开关拨至“DC A”范围内的合适挡位（被测电流超过 200 mA 时应拨至 20 mA/10 A 挡）。

（3）红色表笔接“mA”插孔（<200 mA）或“10A”插孔（>200 mA），黑色表笔接“COM”插孔，表笔与被测电路串联。

5. 交流电流的测量

（1）交流电流有四挡，分别为：200 μA、2 mA、20 mA、200 mA。

（2）把量程开关拨至“AC A”范围内的合适挡位，表笔接法同直流电流的测量。

6. 电阻的测量

（1）电阻有六挡，分别为：200 Ω、2 kΩ、20 kΩ、200 kΩ、2 MΩ、20 MΩ。

（2）将量程开关拨至“Ω”范围内的合适挡。红色表笔接“V · Ω”插孔。

（3）200 Ω 挡的最大开路电压约为 1.5 V，其余电阻挡约为 0.75 V。

（4）电阻挡的最大允许输入电压为 250 V（DC 或 AC）是指误用电阻挡测量电压时仪表的安全值，决不是表示可以带电测量电阻。

7. 二极管的测量

（1）将量程开关拨至二极管挡。

（2）红色表笔插入“V · Ω”插孔，接二极管正极；黑表笔插入“COM”插孔，接二极管负极。此时为正向测量，若管子正常，则测锗管时应显示为 0.150 ~ 0.300 V，测硅管时显示为 0.550 ~ 0.700 V。

（3）进行反向测试时，二极管的接法与上相反，若管子正常，将显示“1”；若管子不正常，则显示“000”。

8. 检查线路通断

（1）检查线路通断（蜂鸣器），将量程开关旋至蜂鸣器挡，红、黑色表笔分别接“V · Ω”和“COM”。

（2）若被测线路电阻低于规定值（20±10）Ω，蜂鸣器可发出声音，表示线路是连通的。利用蜂鸣器来检查线路通断既迅速又方便，因为使用者不需读出电阻值，仅凭听觉即可作出判断。

七、万用表使用实例

（一）检测电位器

使用万用表检测电位器的方法如下：

（1）用万用表欧姆挡测量电位器的两个固定端的电阻，并将测量出来的电阻与标称值进行比较。若万用表测量出来的阻值比标称值大得多，则电位器已损坏；若指示的数值不稳定，不停跳动，则表明电位器内部接触不好。

（2）用万用表欧姆挡测量电位器滑动端与固定端的阻值变化情况。移动滑动端，若万用表测量出来的阻值从小到大连续变化，最小值越小，最大值越接近标称值，则表明电位器质量越好；若测量出来的阻值间断或不连续，则说明电位器滑动端接触不良，不能选用。

（二）检测电容器

使用万用表对电容器的容量、漏电、极性、电容器是否击穿进行初步测量和检查，具体检查方法如下：

1. 固定电容器的检测

（1）检测 10 pF 以下的小电容。因为 10 pF 以下的固定电容器容量很小，用万用表进行测量，只能定性地检查其是否有漏电、内部短路或击穿现象。测量时，万用表一般选用 $R\times10$ kΩ 挡，用两表笔分别任意接电容的两个引脚，阻值应为无穷大。若阻值为零，则说明电容漏电损坏或内部击穿。

（2）检测 10 pF ~ 0.01 μF 固定电容器，可根据其是否有充电现象，进而判断好坏。万用表选用 $R\times1$ kΩ 挡。先用万用表两表笔任意接触电容器两个引脚，再调换表笔触碰电容器两个引脚，如果电容器性能良好，则万用表指针会向右摆动一下，随即迅速向左回转，返回无穷大位置。应注意的是：在测试操作时，特别是在测量小容量电容时，要反复调换被测电容器的两个引脚，才能明显地看到万用表指针的摆动。

（3）对于 0.01 μF 以上的固定电容，可用万用表的 $R\times10$ kΩ 挡直接测试电容器有无充电过程以及有无内部短路或漏电，并可根据指针向右摆动幅度估计出电容器的容量。

2. 电解电容器的检测

（1）万用表量程选择。因为电解电容的容量较一般比固定电容大得多，所以测量时，应针对不同容量选用合适的量程。一般情况下，1 ~ 47μF 的电容可用 $R\times1$ kΩ 挡测量，大于 47 μF 的电容可用 $R\times100$ Ω 挡测量。

（2）性能判别。将万用表红表笔接负极，黑表笔接正极，在刚接触的瞬间，万用表指针即向右偏转较大偏度（对于同一电阻挡，容量越大，摆幅越大），接着逐渐向左回转，直到停在某一位置。此时的阻值便是电解电容的正向漏电阻，此值略大于反向漏电阻。实际使用经验表明，电解电容的漏电阻一般应在几百千欧以上，否则，电容将不能正常工作。在测试中，若正向、反向均无充电的现象，即表针不动，则说明容量消失或内部断路；若阻值很小或为零，说明电容漏电大或已击穿损坏，不能再使用。

（3）极性判别。对于正、负极标志不明的电解电容器，可利用上述测量漏电阻的方法加以判别。即先任意测一下漏电阻，记住其大小，然后交换表笔再测出一个阻值。两次测量中阻值大的那一次便是正向接法，即黑表笔接的是正极，红表笔接的是负极。

（4）容量估测。使用万用表电阻挡，采用给电解电容进行正、反向充电的方法，根据指针右摆幅度大小，可估测出电解电容的容量。电容器的测试如表 2-3 所示。

3. 可变电容器的检测

（1）用手轻轻旋动转轴，应感觉十分平滑，不应感觉时松时紧甚至有卡滞现象。将转轴向前、后、左、右、上、下等各个方向推动时，转轴不应有松动的现象。

（2）用一只手旋动转轴，另一只手轻摸动片组外缘，不应感觉有任何松脱现象。转轴与动片之间接触不良的可变电容器，不能继续使用。

（3）将万用表置于 $R\times10$ kΩ 挡，一只手将两个表笔分别接可变电容器的动片和定片的引出端，另一只手将转轴缓缓旋动几个来回，万用表指针都应在无穷大位置不动。在旋动转轴的过程中，如果指针有时指向零，说明动片和定片之间存在短路点；如果碰到某一角度，万用表读数不为无穷大，而是出现一定阻值，说明可变电容器动片与定片之间存在漏电现象。

表 2-3　电容器的测试

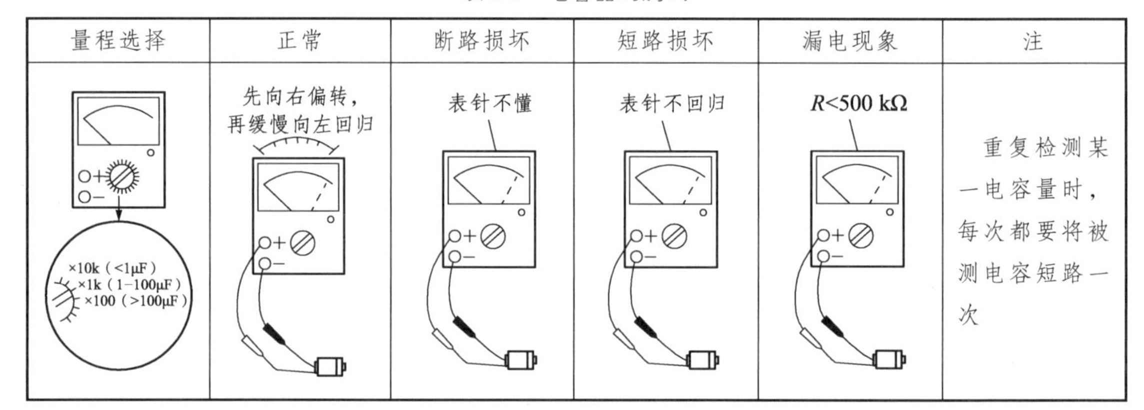

量程选择	正常	断路损坏	短路损坏	漏电现象	注
×10k（<1μF） ×1k（1–100μF） ×100（>100μF）	先向右偏转，再缓慢向左回归	表针不懂	表针不回归	R<500 kΩ	重复检测某一电容量时，每次都要将被测电容短路一次

（三）检测二极管

使用万用表检测二极管，主要是判断其极性和质量好坏。

1. 判别极性

用万用表判断二极管的极性。根据二极管正向电阻小、反向电阻大的特点，将万用表拨到电阻挡。小功率二极管一般用 $R\times100$ Ω 或 $R\times1$ kΩ 挡，不能用 $R\times1$ Ω 挡，因其电流较大，可能损坏二极管，也不能用 $R\times10$ kΩ 挡，因为 $R\times10$ kΩ 电压过高，可能击穿管子；大功率二极管可选用 $R\times1$ Ω 挡。将两表笔分别接触二极管两个电极，测得一个电阻值，交换电极再测一次，从而得到两个电阻值。一般来说正向电阻小于 5 kΩ，反向电阻大于 500 kΩ，测量方法如图 2-15 所示。性能好的二极管，一般反向电阻比正向电阻大几百倍。阻值较小的一次，与黑表笔相接的一端为阳极；阻值较大的一次，与黑表笔相接的一端为二极管的阴极。

2. 质量好坏的判别

将万用表拨到电阻挡（一般用 $R\times100$ Ω 或挡或 $R\times1$ kΩ 挡），如果测得的正、反向电阻很小或等于零，则说明管子内部已击穿或短路；如果正、反向电阻均很大或接近无穷大，说明管子内部已开路；如果电阻值相差不大，说明管子性能变差。上述三种情况发生时，二极管均不能使用。

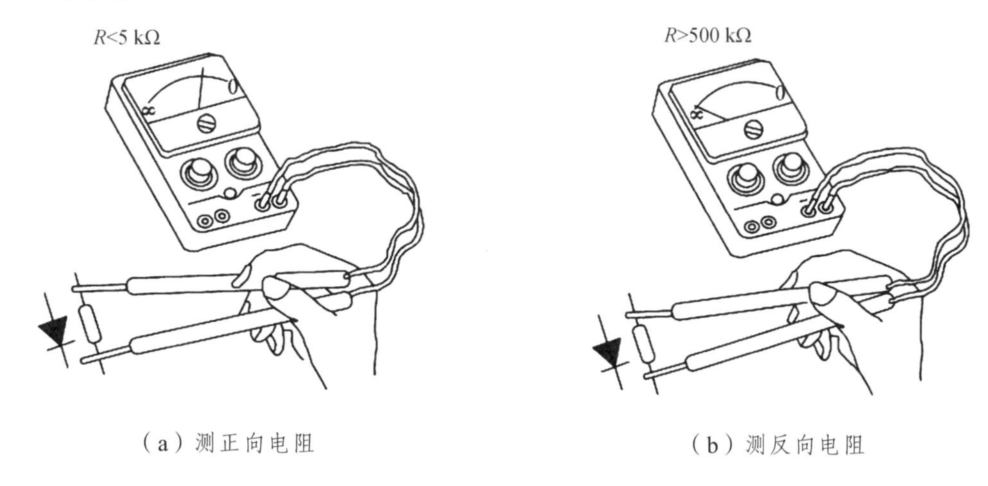

（a）测正向电阻　　（b）测反向电阻

图 2-15　二极管的极性判断

（四）检测三极管

使用万用表检测三极管，主要包括三极管管脚极性确定和质量好坏判断。将万用表置于 $R\times1$ kΩ 挡，测量方法如下：

1. 三极管管脚极性和管型判别

（1）判断基极并确定三极管类型。

先用黑表笔接某一管脚，红笔分别接另外两引脚，测得两个电阻值。再将黑表笔换接另一管脚，重复以上步骤，若测得的两个电阻值都很小，则这时黑表笔所接就是基极 b。该三极管是 NPN 型三极管，测量方法如图 2-16 所示。

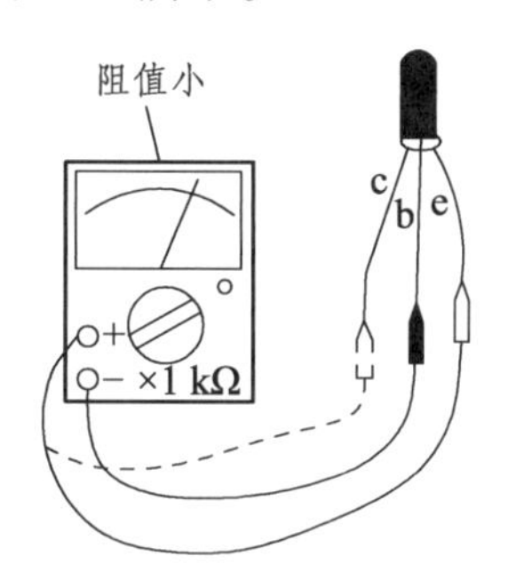

图 2-16　判断三极管基极

若为 PNP 管则应用红表笔与假定的 b 极相接，用黑表笔接另外两个电极。两次测得电阻均很小时，红表笔所接的为 b 极，且可确定三极管为 PNP 管。

（2）确定集电极 c 和发射极 e。

从三极管的结构图上看，发射极 e 和集电极 c 并无大的区别，可以互换使用。实际上，发射区与集电区面积和掺杂浓度有很大的差异。若是 NPN 管，可将黑表笔和红表笔分别接触两个待定的电极，然后用手指捏紧黑表笔和 b 极（相当于接入一个 100 kΩ 的电阻，注意不能将两极短路），观察表的指针摆动幅度，见图 2-17（a）；然后将黑、红表笔对调，按上述方法重测一次。比较两次表针摆动幅度，摆动幅度较大的一次黑表笔所接的管脚为 c 极，红表笔所接的为 e 极，其原理图如图 2-17（b）所示。

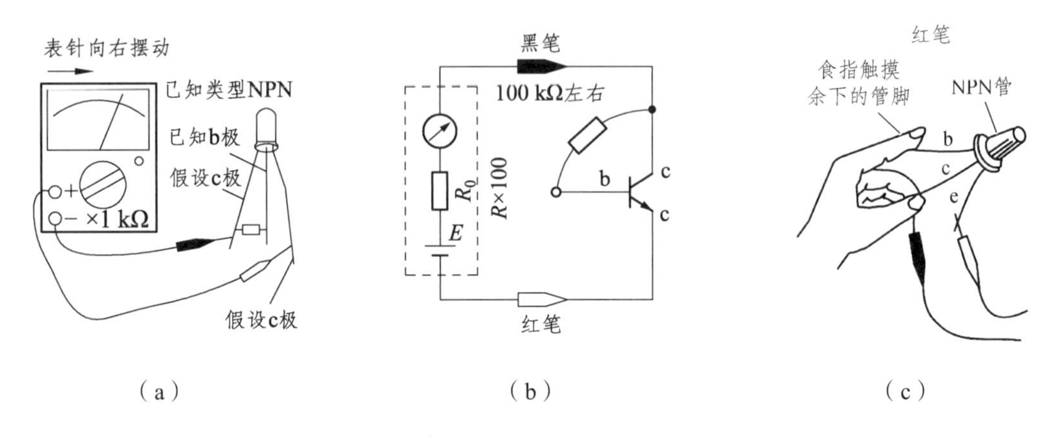

图 2-17　判断 c 极和 e 极

在实际测量中用手指代替基极电阻的做法如图 2-17（c）所示。

若为 PNP 管，则上述方法中将黑、红表笔调换即可。

2. 三极管质量好坏判断

以 NPN 型管为例判断三极管质量好坏判断。用万用表的 $R\times1$ kΩ 挡，将黑表笔接在三极管的基极，红表笔分别接在三极管的发射极和集电极，测得两次的电阻值均为 kΩ 级（若三极管为锗管，则阻值为 1 kΩ 左右；若为硅管，则阻值为 7 kΩ 左右），然后将红表笔接在基极，黑表笔分别接三极管的 e 极和 c 极，若测得的电阻为无穷大，可初步判定三极管完好。然后用万用表测量三极管 e 极和 c 极之间的电阻，测量阻值也应该是无穷大。若测量结果符合上述结论，则三极管是好的。

如果两次测得三极管发射极和集电极之间的电阻都为零或都为无穷大，则说明三极管发射极和集电极之间短路或开路，此三极管已不再可用。

若为 PNP 型三极管，只需对调万用表的红黑表笔即可判断其质量好坏。

（五）万用表使用注意事项

使用万用表时，必须注意以下几点。

（1）测量时应将万用表的红色表笔的连线接到红色的接线柱上或插到标有“+”号的插孔内，黑色表笔应接到黑色接线柱上或插到标有“-”号的插孔内。

（2）使用万用表一般手握表笔操作，注意手不要接触表笔金属部分，以保证安全和测量准确。

（3）使用前必须按被测量的种类和大小将转换开关拨在相应的挡上。

（4）测电流时，仪表应和电路串联；测电压时，仪表应和电路并联。如果对被测量心中无数，应将转换开关拨到最大量限试测，然后再选择合适的量限。转换量限时不要带电操作，以免损坏仪表。测直流电流和电压时，要注意极性。

（5）测量电阻时不能带电测量，被测电阻不能有并联支路。测量电阻前应进行欧姆调零，更换电阻倍率后，应重新调零。

（6）测量的直流电压上叠加有交流或脉冲电压时，应注意万用表的耐压，以免峰值电压过大损坏万用表。

（7）万用表用完后，应将转换开关拨到非测量位置或交流电压的最高挡，以免下次使用时不注意测量种类和量限而损坏万用表。

总之，万用表作为常用测量仪表，尽管简单，但也必须正确使用，耐心操作，才能保证仪表精度，提高测量准确性。

Task Ⅰ Inspection and use of multimeter

Multimeters commonly used in electric power measurement, which are used to measure AC current and voltage, resistance, DC current, etc., are divided into analog multimeters and electronic digital multimeters. The analog multimeter is a moving-coil DC ammeter used to indicate the measured value. All measurements are converted into the driving current of the moving-coil DC ammeter. The deflection angle of the DC ammeter pointer is used to indicate the measured value. In addition, there are shunt (to expand the measuring range of current), multiplier (to expand the measuring range of voltage), rectifier (to turn AC into DC), battery (to provide power supply for measuring resistance), selector switch and so on. Compared with the electronic digital multimeter, the reading accuracy of the pointer gauge of the analog multimeter is poor, but the pointer swing process is more intuitive, and its swing speed and amplitude can sometimes objectively reflect the measured value. On the other hand, the reading of the electronic digital multimeter is intuitive, but the process of digital change looks very complex and difficult to watch.

Fig. 2-1 shows the outline drawing of an analog multimeter, and Fig. 2-2 shows its dial, indicating the functions of each part.

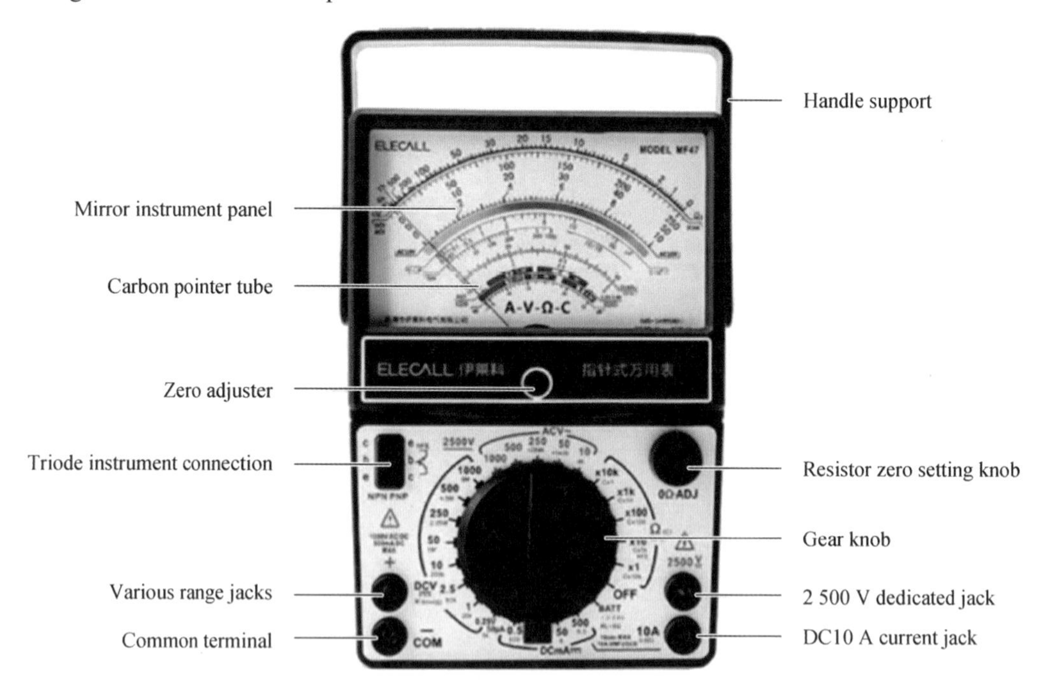

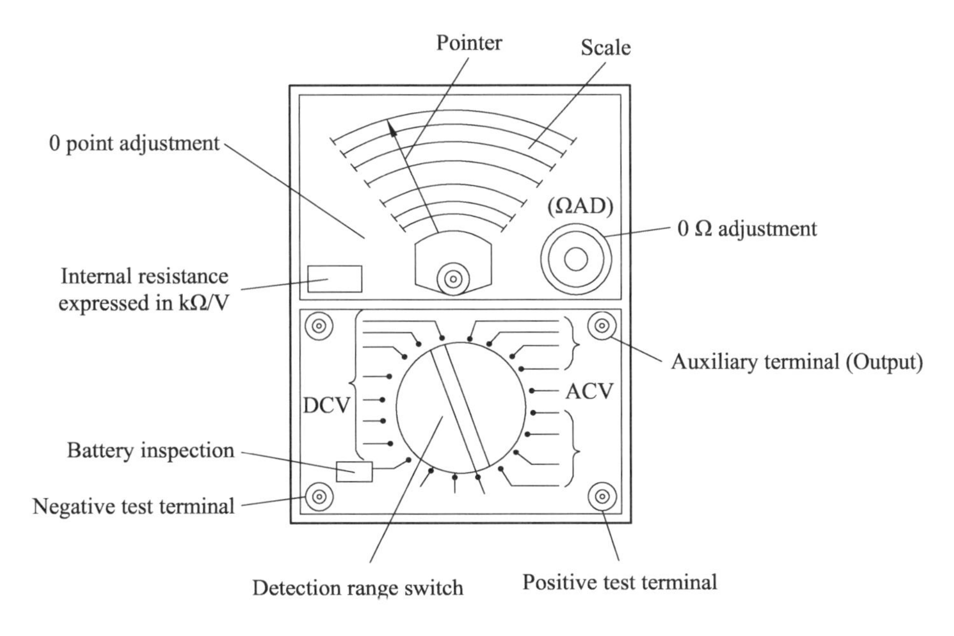

Fig. 2-1 Analog multimeter

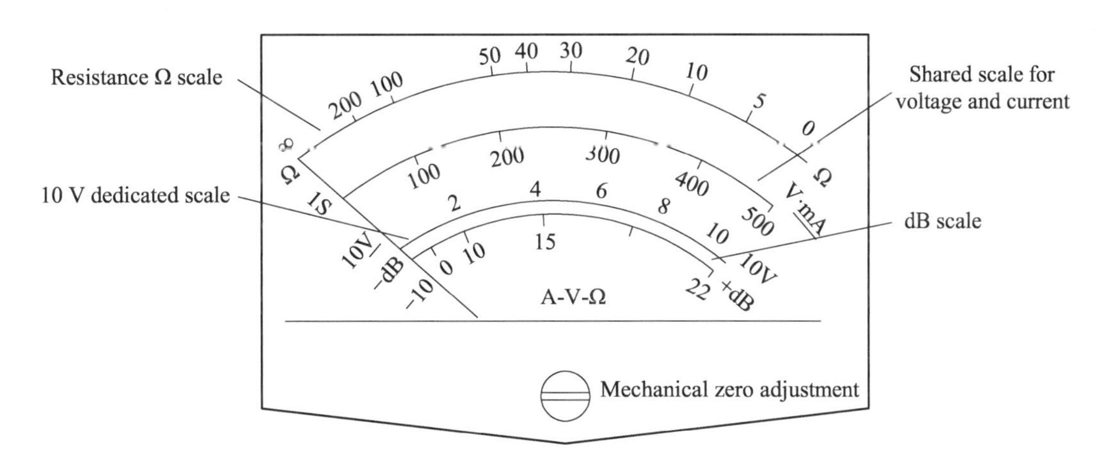

Fig. 2-2 Analog multimeter dial

Ⅰ. Overview of multimeter

1. Functions of the main parts of a multimeter

1) Meter head calibrator

When the multimeter probe is open circuited, the pointer shall point to the 0 position (on the left side of the dial, the 0 value of the voltage and current scale). If it is not at the 0 position, a screwdriver can be used to fine adjust the pointer to the 0 position, which is also known as zero adjustment.

2) Measuring range changeover switch

The current, voltage, resistance or other different measuring ranges are converted by this switch, and the number of each position is represented by the value of the full scale on the dial.

3) Zero ohm adjustment

When measuring the resistance, first short-circuit the two probes. At this time, the probe shall point to 0 (the right side of the dial, the 0 value of the resistance scale). If it is not 0, fine tune this button (potentiometer) to make the needle point to 0.

4) Measuring terminal

There are two probes attached to the multimeter. When measuring, the red probe is inserted into the "+" terminal, and the black probe is inserted into the "−" terminal.

2. Performance of multimeter

The performance of the multimeter is as shown in Tab. 2-1 and Tab. 2-2.

Tab. 2-1 Maximum scale value of a multimeter

Measuring item	Maximum scale value
DC voltage/V	0.25, 1, 2.5, 10, 50, 250, 1000 (internal resistance 20 kΩ/V)
AC voltage/V	1.5, 10, 50, 250, 1000 (internal resistance 20 kΩ/V)
DC current/mA	3 000,30 000,300 000
Audio level/dB	0 ~ +22 (AG10V range)

Tab. 2-2 Error of a multimeter

Measuring item	Tolerance
DC voltage and current	±3% of maximum scale value
AC voltage	±4% of maximum scale value
Resistance	± 3% of the length of the dial

Ⅱ. Measurement of DC current

1. Basic principle of measuring DC current

A magnetoelectric meter head is an ammeter, but its range is I_g (usually several microamperes to dozens of microamperes). To measure a large current, a resistor with an appropriate resistance value can be connected in parallel at both ends of the meter head, as shown in Fig. 2-3. Among them, R_s is called shunt resistor (diverter), and the magnitude of the resistance can be calculated using the following equation:

$$R_s=R_g/(n-1) \tag{2-1}$$

Where, $n=I/I_g$, represents the multiple of the meter head range expansion.

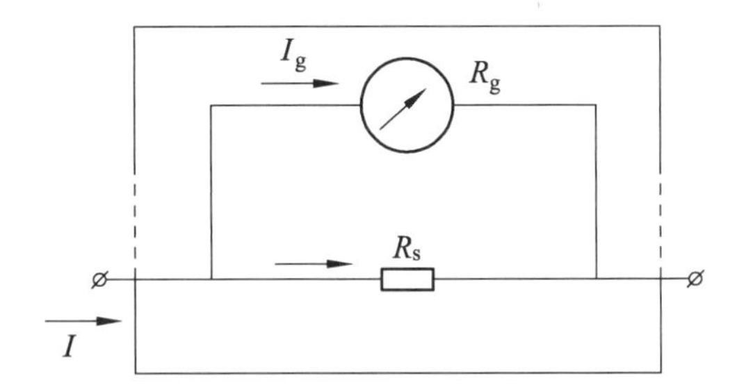

Fig. 2-3 Schematic diagram of single range ammeter

When R_s is a constant value, the measured current I is proportional to the magnitude of the current flowing through the meter head, so the deflection angle of the meter head pointer can reflect the magnitude of the measured current.

The multi-range ammeter, that is, the resistors with different resistance values connected in parallel at both ends of the meter head, is connected to the circuit by the changeover switch. There are two connection methods of shunt: independent split and closed tap. In order to protect the safety of the meter head, if the switch contact is poor, the independent split connection type may damage the meter head, while the closed tap connection type can avoid damage to the meter head, so the closed tap connection type is often used in circuit connection, as shown in Fig. 2-4. This ammeter has three ranges, which are I_1, I_2 and I_3, respectively.

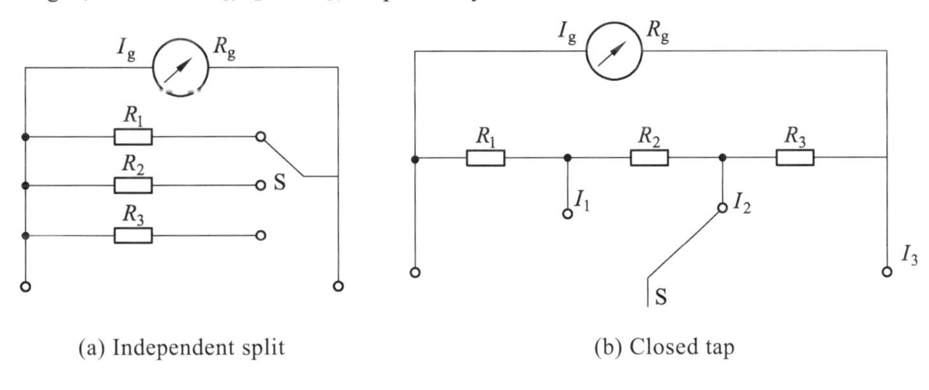

(a) Independent split (b) Closed tap

Fig. 2-4 Schematic diagram of multi-range ammeter

2. Practical operation for measuring DC current

To measure the DC current of a circuit, it is necessary to cut off the circuit of the tested part and connect a multimeter in series with the circuit, as shown in Fig. 2-5.

When measuring, turn the measuring range button to the position of DC mA, turn on the circuit, and then select the measuring range according to the measured value. As shown in Fig. 2-5, when measuring, connect the positive end (red probe) of the multimeter to the high voltage end and the negative end (black probe) of the multimeter to the low voltage end. If the polarity is reversed, the needle will swing in the opposite direction, which may cause a fault of the multimeter. This is something to be aware of.

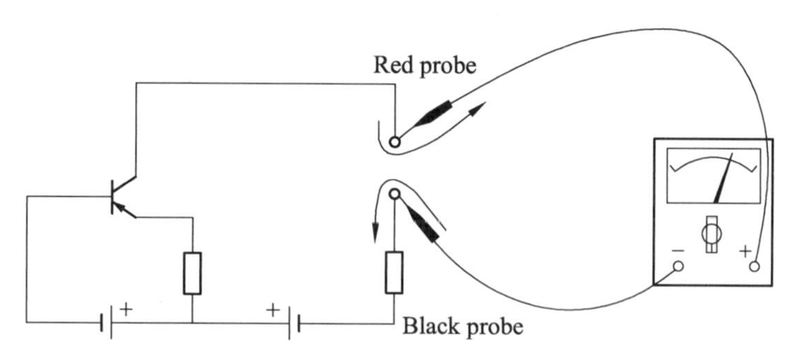

Fig. 2-5 Measurement of DC current

Ⅲ. Measurement of DC voltage

1. Basic principle of measuring DC voltage

DC voltages smaller than U_g ($U_g=I_gR_g$) can be measured with a single magnetoelectric meter head. If a larger voltage needs to be measured, according to the principle that series resistors can be used to divide the voltage, an appropriate resistor can be connected in series on the meter head, as shown in Fig. 2-6.

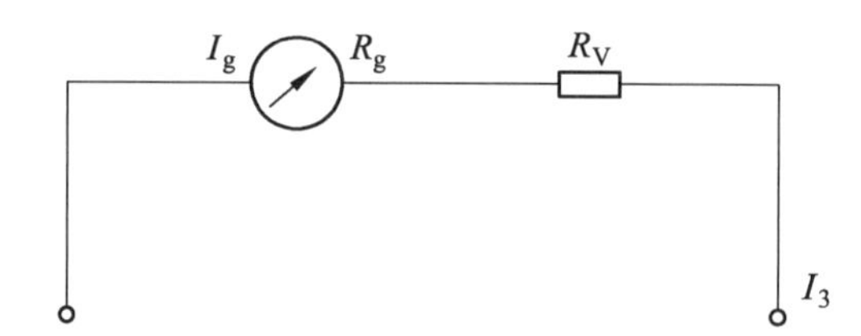

Fig. 2-6 Single range DC voltage measurement

In Fig. 2-6, R_v is the divider resistor, and the resistance value is calculated using the following equation:

$$R_v=(U-I_gR_g)/I_g=(mU_g-I_gR_g)/I_g=(m-1)I_g \qquad (2\text{-}2)$$

Where, $m = U/U_g$, represents the multiple of the meter head range expansion.

When R_v is a constant value, the measured voltage U is proportional to the magnitude of the current flowing through the meter head, so the deflection angle of the meter head pointer can reflect the magnitude of the measured voltage.

If multiple divider resistors are connected in series with the meter head, a multi-range DC voltmeter can be made. The connection mode is divided into single-use type and common type, and the circuit principle is as shown in Fig. 2-7.

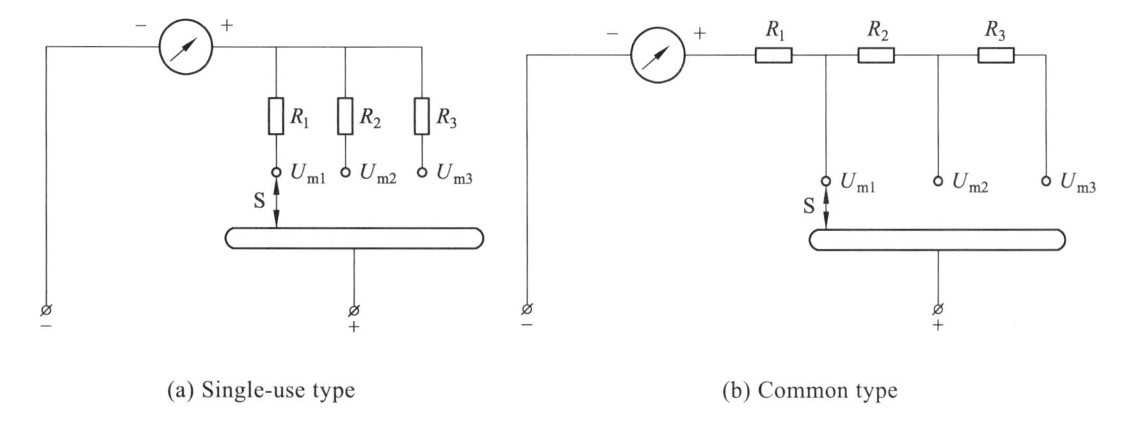

(a) Single-use type (b) Common type

Fig. 2-7 Multi-range DC voltage measurement

2. Practical operation for measuring DC voltage

To measure the DC voltage in the circuit, first set the measurement range switch to the DC voltage position and select the appropriate measurement range. The measurement example is as shown in Fig. 2-8, and the voltage drop on the collector load resistor needs to be measured. Connect the positive end (red probe) of the multimeter to the high voltage end and the negative end (black probe) of the multimeter to the low voltage end. If the polarity is reversed, the needle will swing in the opposite direction, which may cause a fault of the multimeter.

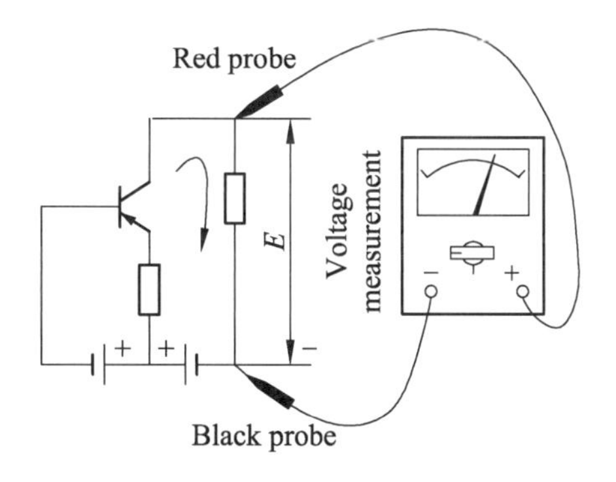

Fig. 2-8 Measurement of DC voltage

Ⅳ. Measurement of AC voltage

1. Basic principle of measuring AC voltage

Because the head of the multimeter is a magnetoelectric measuring mechanism, it can only measure DC. Therefore, when measuring AC, rectification measures must be taken to convert AC power into DC power. There are various ways to rectify AC signals, the most common being average rectification and peak rectification. On the multimeter, the average rectification method is commonly used, and the average rectification can be divided into two types: half-wave rectification and full-wave rectification.

(1) Half-wave rectifier circuit: It is a common circuit that utilizes the unidirectional conductivity of a diode for rectification. The rectification method of removing half cycle and leaving half cycle is called half-wave rectification. As shown in Fig. 2-9, rectifier diode VD_1 is connected in series with the meter head to form a branch, while diode VD_2 is connected in parallel at both ends of the branch connected in series with the meter head and VD_1. Due to the unidirectional conductivity of the diode, the input signal is subjected to the action of VD_1 and VD_2, and the current flowing through the meter head is a unidirectional pulsating current, whose waveform is as shown in Fig. 2-9(a)(b). When VD_2 is on, the forward resistance of the diode is very low, so the voltage at terminals a and b is also very low, usually only 0. 3–0.7 V, which protects the VD_1 from being broken down by the reverse voltage.

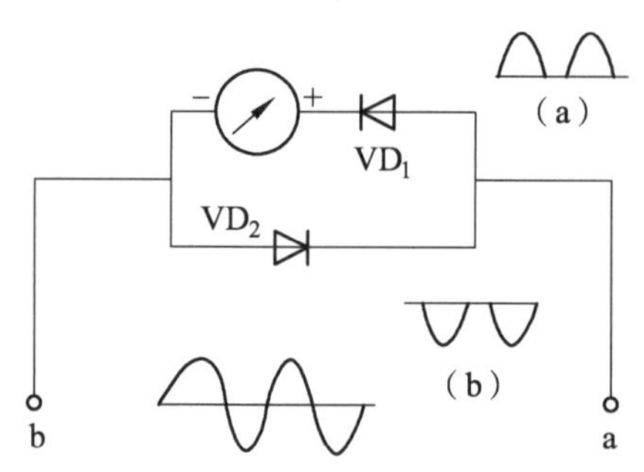

Fig. 2-9 Half-wave rectifier circuit

(2) Full-wave rectifier circuit: As shown in Fig. 2-10, it is a bridge rectifier circuit composed of four rectifier diodes, which become the four arms of the bridge. One of the two diagonals is connected to the AC power supply and the other to the magnetoelectric measuring mechanism. Due to the action of diodes VD_1, VD_3, VD_2 and VD_4, two half-wave currents in the same direction flow through the meter head in one cycle of AC voltage, and their waveforms are as shown in Fig. 2-10(a). If the applied AC voltage is equal, the current flowing through the meter head in the full-wave rectifier circuit is twice as large as that in the half-wave rectifier circuit. Therefore, the full-wave rectifier circuit has higher sensitivity than the half-wave rectifier circuit, or its rectification efficiency is twice as high.

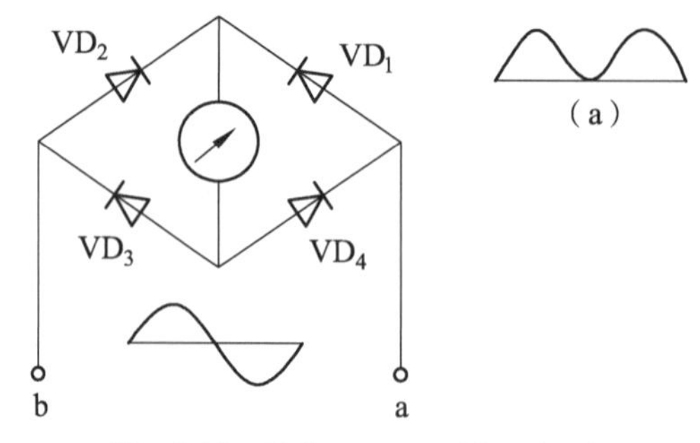

Fig. 2-10 Full-wave rectifier circuit

(3) Multi-range AC voltmeter: The meter head circuit with rectifier circuit is connected in series with additional resistors of various values to form a multi-range AC voltmeter. Similar to the

measurement circuit for DC voltage range, the measurement circuit for multi-range AC voltage range is also divided into two types: single-use type and common type, as shown in Fig. 2-11.

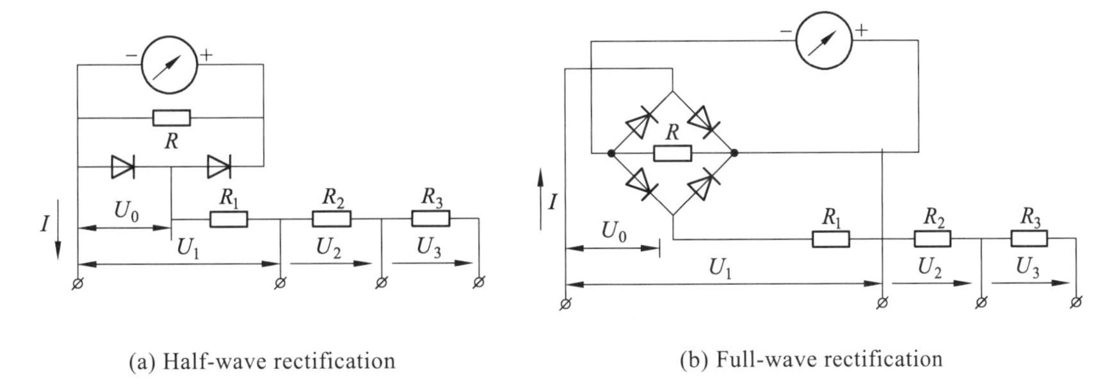

(a) Half-wave rectification　　(b) Full-wave rectification

Fig. 2-11　Rectified AC voltmeter circuit

2. Practical operation for measuring AC voltage

The measurement of the tap voltage of the power supply transformer, the inspection of the display filament voltage, and the inspection of the AC 220 V voltage are all covered in the measurement range of AC voltage. When measuring AC voltage, turn the measuring range switch to the AC position and further select the measuring range. The polarity of the probe is arbitrary.

To measure the AC component superimposed on the DC voltage, a 0.1 μF capacitor can be connected in series on the probe to isolate the DC component, and some multimeters have built-in capacitors. Generally speaking, it is not possible to measure AC signals above 50 kHz.

Ⅴ. Measurement of DC resistance

1. Basic principle of measuring DC resistance

The basic circuit of the resistance gear of the multimeter is as shown in Fig. 2-12.

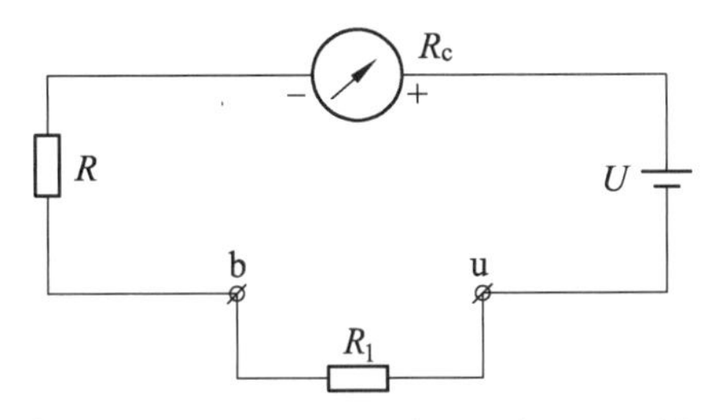

Fig. 2-12　Schematic circuit diagram of measuring resistance with a multimeter ohm gear

According to Ohm's law, the current flowing through the measured resistor is:

$$I = \frac{U}{R_c + R + R_x} \qquad (2\text{-}3)$$

From the above, it can be seen that when the battery voltage U, fixed resistance R, and R_c remain constant, the magnitude of the current flowing through the meter head corresponds one-to-one to the magnitude of the measured resistance R_x. Therefore, the deflection angle of the meter head pointer can be used to reflect the magnitude of the measured resistance. At the same time, the following conclusions can be drawn.

(1) According to the principle of resistance measurement, the relationship between the current flowing through the meter head and the measured resistance is not linear, therefore, the scale of the ohmmeter is uneven. When R_x is infinite, I=0, the deflection angle of the pointer is 0; When the measured resistance R_x = 0, the current I flowing through the meter head happens to be the full bias current I_c of the meter head, and at this time, the pointer deflects at full scale. It can be seen that the ohm gear scale is a reverse scale, which is exactly opposite to the scale direction of its current and voltage scales, as shown in Fig. 2-13.

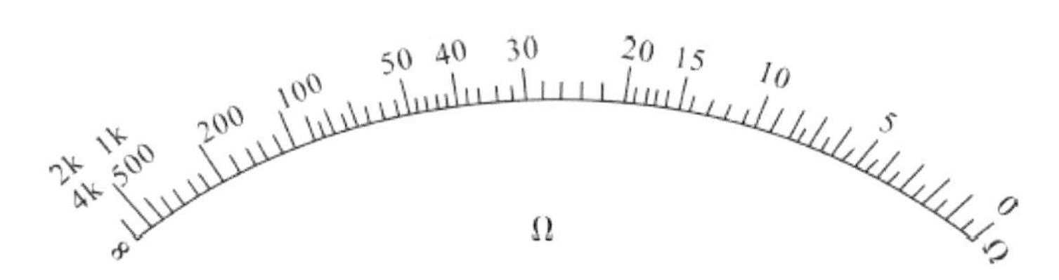

Fig. 2-13 Ohm gear scale

(2) With the increase of the use or storage time of the multimeter ohm gear, the battery terminal voltage will gradually decrease, which will inevitably reduce the working current during measurement, resulting in measurement errors. The most obvious error is that when R_x =0, the head pointer cannot reach full scale deflection, that is, it cannot reach 0 Ω scale line. Therefore, the actual multimeter ohm gear is equipped with zero ohm regulator, and the commonly used 0 Ω regulator generally uses a voltage divider circuit. Adjust the shunt current by adjusting the potentiometer, so as to ensure that when R_x =0, the current flowing through the meter head is equal to the full bias current of the meter head, so that the pointer reaches the 0 Ω scale line.

(3) As shown in Fig. 2-12, when R_x =R_c+R, there are:

$$I = \frac{U}{R_c + R + R_x} = \frac{U}{2(R_c + R)} = \frac{I_c}{2} \tag{2-4}$$

Where, I_c is the full bias current of the meter head. When the pointer of the meter head is at the center of the scale, the indicated value is called the ohmic central value. The ohmic value it indicates is exactly the total internal resistance value of the limit. First of all, because the scale of the ohmmeter is uneven, knowing the ohmic central value, the effective measuring range of the ohmmeter is determined, so that the effective measuring range is generally ($\frac{1}{10}$ ~ 10) times the ohmic central value. If the measured resistance is too large beyond this range, the measuring range needs to be changed. Secondly, according to the ohmic central value, the measuring range can be expanded by Decimal multiples. This makes it possible to share a single scale for all ranges, making it easy to read.

(4) With the expansion of the measuring range of the resistance, the total internal resistance of the ohmmeter increases, and the working current of the meter head decreases accordingly. When R_x =0,

the pointer can not point to the zero scale. To solve this problem, firstly, while keeping the battery voltage constant, change the shunt resistance of the measuring circuit to adapt to the requirements for working current at different ranges. Second, with the increase of the working voltage, the internal resistance of the meter increases when the resistance is high, but when the battery voltage is increased, the current of the meter can still reach full bias when R_x =0, and the higher battery voltage can be connected through the changeover switch.

2. Practical operation for measuring DC resistance

When measuring the resistance value, first turn the measuring range switch to the Ω position, and then select the measuring range based on the resistance value. When measuring resistance, first short-circuit the two probes so that the needle points at 0 Ω, and then measure the actual resistance. The polarity of the probe is arbitrary. To measure the resistance in a circuit with large capacitance, the charging charge on the capacitor must be discharged before measurement.

Ⅵ. Electronic digital multimeter

1. Panel structure of electronic digital multimeter

The panel structure of an electronic digital multimeter mainly includes a LCD display screen, power supply switch, functional range switch, h_{FE} jack, and input jack. Take the DT-830 electronic digital multimeter as an example, its panel is as shown in Fig. 2-14.

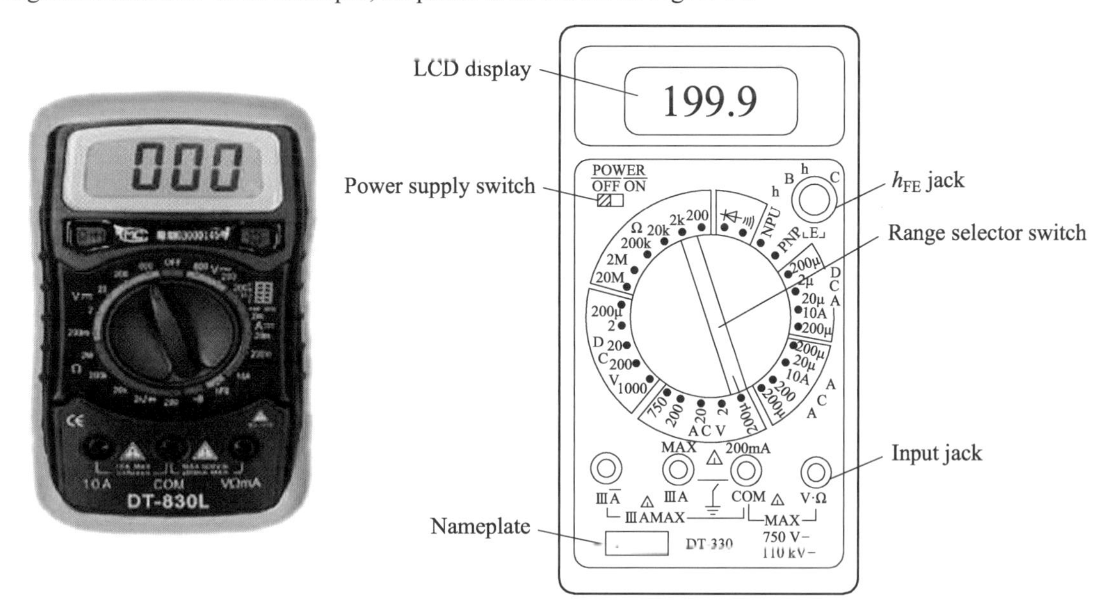

Fig. 2-14 DT-830 electronic digital multimeter

The DT-830 electronic digital multimeter panel consists of five parts, and the names and functions of all parts are as follows:

(1) LCD display screen: displays a variety of measured values, including decimal point, plus or minus sign and overflow status.

(2) Power supply switch: turn on and off the battery power supply in the meter.

(3) Functional range switch: convert different measurement functions and ranges according to specific situations.

(4) h_{FE} jack: used for measuring triode parameters.

(5) Input socket: used for externally connecting testing probes.

2. Main functions and technical indicators of electronic digital multimeter

1) Main functions

Electronic digital multimeter can not only measure DC current and voltage, AC current and voltage, and resistance like analog multimeter, but also measure diode junction voltage, triode transistor h_{FE}, and line on/off.

2) Technical indicators

(1) Measurement accuracy: the errors of electronic digital multimeter mainly include reference source errors, input amplifier errors, nonlinear errors, and quantization errors.

(2) Resolution: refers to the numerical value corresponding to the last digit.

(3) Input impedance: the equivalent impedance of the input circuit seen from the input terminal when it is in the working state.

(4) Measuring speed and response time: the measuring speed refers to the number of times the measurement is completed according to the prescribed accuracy in a unit time. Response time refers to the time interval between the moment of sudden change of the input signal and the new stable display value that meets the accuracy.

3. Usage of electronic digital multimeter

1) Panel

(1) The display screen is a large font LCD display screen. The instrument has the function of automatic zero adjustment and automatic polarity display. If the polarity of the measured voltage or current is negative, a negative sign "–" will appear before the displayed value. When the voltage of the laminated battery is less than 7 V, a low voltage indicator is displayed on the upper left of the display screen, indicating that the battery needs to be replaced. "1" or " –1" is displayed when exceeding the range, depending on the polarity of the electric quantity being measured. The decimal point is synchronously controlled by the range switch to move the decimal point to the left or right.

(2) Power supply switch. The symbols "OFF" and "ON" are marked below POWER. Turn the power supply switch to "ON" and turn on the power supply to use the instrument; After use, the switch should be turned to the "OFF" position to avoid battery depletion.

(3) The functional range switch can complete the selection of test function and range.

(4) The h_{FE} jack is a four-core socket, marked with B, C, and E. There are two E sockets connected internally. When measuring the h_{FE} value of the triode transistor, the three electrodes shall be inserted into the B, C and E sockets, respectively.

(5) There are four input jacks marked with "10 A", "mA", "COM", and "V • Ω", respectively. The words "MAX 750 V ~ " and "1,000 V - " are marked between " V • Ω" and "COM", indicating that the AC voltage input from these two jacks must not exceed 750 V and the DC voltage must not

exceed 1,000 V. In addition, "MAX 200 mA" is marked between "mA" and "COM", and "MAX 10 A" is also marked between "10 A" and "COM", indicating the maximum allowable values of input AC and DC currents, respectively.

2) Measurement of DC voltage

(1) There are five levels of DC voltage, namely 200 mV, 2 V, 20 V, 200 V, and 1,000 V.

(2) Turn the power supply switch to "ON" and the range switch to the appropriate gear within the "DCV" range.

(3) The red probe is connected to the " V • Ω" jack, the black probe is connected to the "COM" jack, and the probe is connected in parallel with the circuit under test.

(4) Maximum allowable input voltage: DC 1,000 V (200 mV, 2 V, and 20 V ranges); DC 1,100 V (200 V, 1,000 V ranges).

3) Measurement of AC voltage

(1) There are five levels of AC voltage, namely 200 mV, 2 V, 20 V, 200 V, and 750 V.

(2) Turn the range switch to "ACV", select the appropriate gear within the range, and the probe connection method is the same as the measurement of DC voltage.

(3) The required frequency of the measured voltage is 45 ~ 500 Hz, and the maximum allowable input voltage is 750 V (effective value).

4) Measurement of DC current

(1) There are four levels of DC current, namely 200 μA, 2 mA, 20 mA, and 200 mA.

(2) Turn the range switch to the appropriate gear within the "DCA" range (when the measured current exceeds 200 mA, it shall be turned to the 20 mA/ 10 A gear).

(3) The red probe is connected to the "mA" jack (<200 mA) or the "10 A" jack (>200 mA), the black probe is connected to the "COM" jack, and the probe is connected in series with the circuit under test.

5) Measurement of AC current

(1) There are four levels of AC current, namely 200 μA, 2 mA, 20 mA, and 200 mA.

(2) Turn the range switch to the appropriate gear within the "ACA" range, and the probe connection method is the same as the measurement of DC current.

6) Measurement of resistance

(1) There are six levels of resistance, namely 200 Ω, 2 kΩ, 20 kΩ, 200 kΩ, 2 MΩ, and 20 MΩ.

(2) Turn the range switch to the appropriate gear within the "Ω" range. The red probe is connected to the "V • Ω" jack.

(3) The maximum open circuit voltage in the 200 Ω gear is approximately 1.5 V, and that of the remaining resistance gears are approximately 0.75 V.

(4) The maximum allowable input voltage of the resistance gear is 250 V (DC or AC), which refers to the safety value of the instrument when mistakenly using resistance gear to measure voltage, and it does not mean that the resistance can be measured with electricity.

7) Measurement of diode

(1) Turn the range switch to the diode gear.

(2) Insert the red probe into the " V · Ω" jack and connect it to the positive pole of the diode; Insert the black probe into the "COM" jack and connect it to the negative pole of the diode. At this time, it is a forward measurement. If the diode is normal, it shall display as 0.150–0. 300 V when measuring the wrong diode, and display as 0.550 –0.700 V when measuring the silicon tube.

(3) When conducting reverse test, the connection of the diode is opposite to that above. If the diode is normal, it will display "1"; If the diode is not normal, it will display "000".

8) Check that the line is connected or disconnected

(1) To check that the line is connected or disconnected (buzzer), turn the range switch to the buzzer gear, and connect the red and black probes to "V · Ω"and "COM", respectively.

(2) If the resistance of the tested line is lower than the specified value (20 ±10) Ω, the buzzer can make a sound, indicating that the line is connected. Using the buzzer to check the circuit being connected or disconnected is fast and convenient, because the user does not need to read the resistance value and can make a judgment only by hearing.

Ⅶ. An example of using multimeter

1. Testing potentiometer

Use a multimeter to test the potentiometer as follows:

(1) Use the multimeter ohm gear to measure the resistance of the two fixed ends of the potentiometer, and compare the measured resistance with the nominal value. If the resistance measured by the multimeter is much larger than the nominal value, the potentiometer is damaged; If the indicated value is unstable and keeps bouncing, it indicates that the internal contact of the potentiometer is poor.

(2) Use the multimeter ohm gear to measure the resistance change between the sliding end and the fixed end of the potentiometer. When moving the sliding end, if the resistance measured by the multimeter changes continuously from small to large, the smaller the minimum value is, the closer the maximum value is to the nominal value, which indicates that the better the quality of the potentiometer is. If the measured resistance value is intermittent or discontinuous, it means that the contact of the sliding end of the potentiometer is bad and can not be selected. is bad and can not be selected.

2. Testing capacitor

Use the multimeter to measure and check the capacity, leakage, polarity and breakdown of the capacitor. The specific inspection methods are as follows:

1) Test of fixed capacitor

(1) Detect small capacitors below 10 pF. Because the capacity of the fixed capacitor below 10 pF is very small, the measurement with the multimeter can only qualitatively check whether there is leakage, internal short circuit or breakdown. When measuring, the $R\times 10$ kΩ gear of the multimeter is generally selected, and two probes are used to connect any two leads of the capacitor, and the resistance value shall be infinite. If the resistance value is zero, it means that the capacitor is

damaged by leakage or internal breakdown.

(2) For the test of 10 pF ~ 0.01 μF fixed capacitor, it can be judged according to whether it has the phenomenon of charging or not. Select the $R\times1$ kΩ gear of the multimeter. First, touch the two leads of the capacitor arbitrarily with the two probes of the multimeter, and then swap the probes to touch both leads of the capacitor. If the performance of the capacitor is good, the multimeter pointer will swing to the right and then quickly turn to the left to return to the infinite position. It should be noted that during test, especially when measuring small-capacity capacitors, it is necessary to repeatedly swap the two leads of the tested capacitor in order to clearly see the swing of the multimeter pointer.

(3) For fixed capacitors above 0.01 μF, the $R \times 10$ kΩ block gear of the multimeter can be used to directly test the charging process and internal short circuit or leakage of the capacitor, and the capacity of the capacitor can be estimated according to the swing amplitude of the pointer to the right.

2) Test of electrolytic capacitor

(1) Multimeter range selection. Because the capacity of electrolytic capacitor is much larger than that of fixed capacitor, the appropriate range shall be selected for different capacity. In general, the capacitor of 1–47 μF can be measured with $R\times1$ kΩ gear, and the capacitor greater than 47 μF can be measured with $R\times100$ Ω gear.

(2) Performance evaluation. Connect the red probe of the multimeter to the negative pole and the black probe to the positive pole. At the moment of contact, the pointer of the multimeter will deflect to the right by a large degree (for the same resistance gear, the larger the capacity, the greater the swing), and then gradually rotate to the left until it stops at a certain position. At this time, the resistance value is the forward leakage resistance of the electrolytic capacitor, which is slightly larger than the reverse leakage resistance. Practical experience shows that the leakage resistance of the electrolytic capacitor shall generally be above a few hundred kΩ, otherwise, the capacitor will not work properly. In the test, if there is no charging in both the forward direction and the reverse direction, that is, the needle hand does not move, it means that the capacity disappears or there is an internal open circuit; If the resistance value is very small or zero, it means that the capacitor has large leakage or breakdown damage and can no longer be used.

(3) Polarity determination. For Electrolytic capacitors with unknown positive and negative pole signs, the above method of measuring leakage resistance can be used for determination. That is, first arbitrarily measure the leakage resistance, remember the value, and then swap the probe to measure a resistance value. The larger resistance measured in the two measurements is subject to the forward connection method, where the black probe is connected to the positive pole and the red probe is connected to the negative pole.

(4) Capacity estimation. The capacitance of the electrolytic capacitor can be estimated according to the amplitude of the right swing amplitude of the pointer by using the resistance gear of the multimeter and the method of charging the electrolytic capacitor in the forward and reverse directions. The test of the capacitor is as shown in Tab. 2-3.

Tab. 2-3 Testing of capacitor

Range selection	Normal	Open circuit damage	Short circuit damage	Leakage	Note
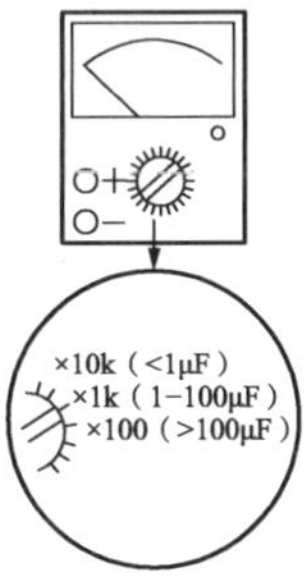	First deflect to the right, then slowly return to the left 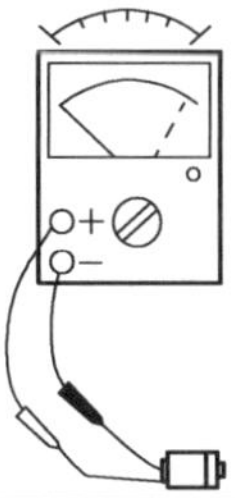	The needle does not move 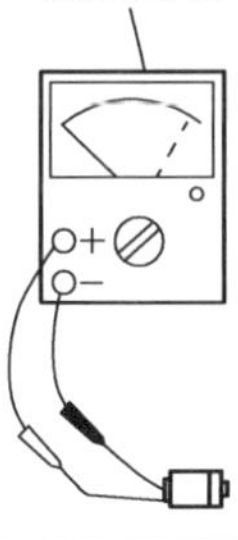	The needle does not return 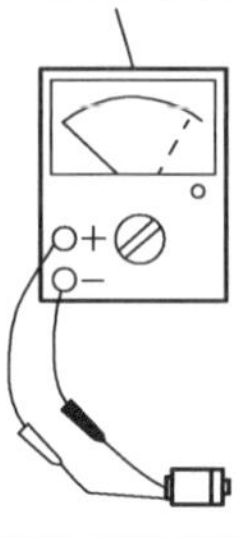	R<500 kΩ 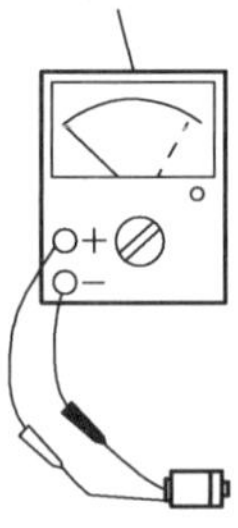	When repeatedly testing a capacitor, the tested capacitance will be short-circuited each time

3) Test of variable capacitor

(1) Gently rotate the pivot with hand, the feeling shall be very smooth and shall not be sometimes loose and sometimes tight, or even stuck. When pushing the pivot forward, backward, left, right, upward, and downward, the pivot shall not be loose.

(2) Rotate the pivot with one hand and gently touch the outer edge of the rotor plate group with the other hand, there shall be no feeling of any looseness. The variable capacitor with poor contact between the pivot and the rotor plate can not continue to be used.

(3) Place the multimeter in the $R\times 10$ kΩ gear, with one hand connecting the two probes to the rotor plate and stator plate leading-out terminals of the variable capacitor, and the other hand slowly rotating the pivot back and forth several times. The multimeter pointer shall remain stationary at the infinite position. In the process of rotating the pivot, if the pointer sometimes points to zero, it means that there is a short circuit point between the rotor plate and the stator plate. If the reading of the multimeter at a certain angle is not infinite but a certain resistance value appears, it indicates that there is leakage between the rotor plate and stator plate of the variable capacitor.

3. Test of diode

The main purpose of using a multimeter to test a diode is to determine its polarity and quality.

1) Polarity determination

Use a multimeter to determine the polarity of the diode. According to the characteristics of low forward resistance and large reverse resistance of the diode, the multimeter is turned on to the resistance gear. Low-power diodes generally use $R\times 100$ Ω or $R\times 1$ kΩ gear, but cannot use $R\times 1$ Ω gear. Due to its high current, the diode may be damaged, and $R\times 10$ kΩ gear cannot be used because the voltage of $R\times 10$ kΩ is too high, which may break through the diode; $R\times 1$ Ω gear can be used for high-power diodes. Contact the two probes separately with the two electrodes of the diode to measure a resistance value. Swap the electrodes and measure again to obtain two resistance values. Generally speaking, the forward resistance is less than 5 kΩ and the reverse resistance is more than 500 kΩ. The measuring method is as shown in Fig. 2-15. For diodes with good performance, the reverse resistance is hundreds of times larger than the forward resistance. For a measurement with a small resistance, the end connected to the black probe is the anode. For a measurement with a large resistance, the end connected to the black probe is the cathode of the diode.

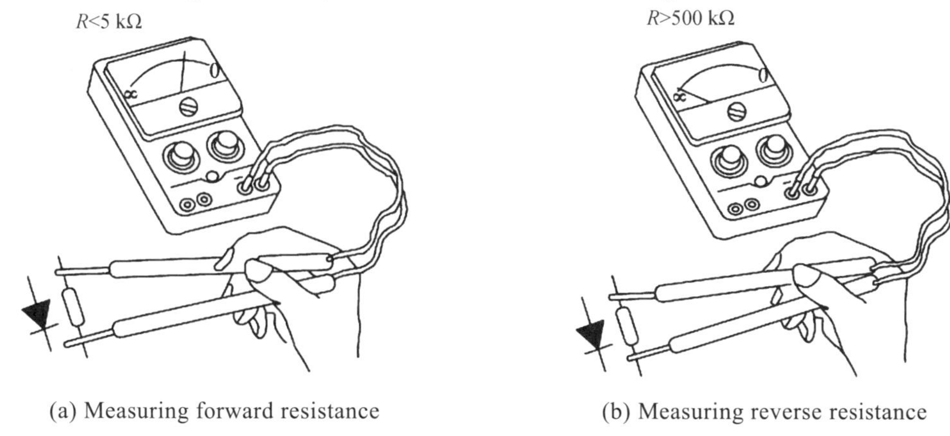

(a) Measuring forward resistance (b) Measuring reverse resistance

Fig. 2-15 Polarity determination of diode

2) Quality determination

Turn the multimeter to the resistance gear (usually $R\times100\ \Omega$ gear or $R\times1\ \text{k}\Omega$ gear). If the measured forward and reverse resistance is very small or equal to zero, it means that the diode has been broken down or short-circuited. If the forward and reverse resistance is very large or close to infinity, it means that the diode is subject to open circuit internally. If the resistance value is not much different, it means that the performance of the diode deteriorates. When any of the above three situations occur, the diode cannot be used.

4. Test of triode

The use of multimeter to test the triode mainly includes the determination of the polarity of the triode pin and the judgment of the quality. Place the multimeter in $R\times1\ \text{k}\Omega$ gear and measure it as follows:

1) Determination of triode pin polarity and type

(1) Determine the base and determine the type of triode.

First, use the black probe to connect one pin and a red probe to connect the other two leads to measure the two resistance values. Then use the black probe to connect another pin, repeat the above steps, until the two resistance values are very small, then the black probe is connected to the base b. This triode is a NPN triode, and the measuring method is as shown in Fig. 2-16.

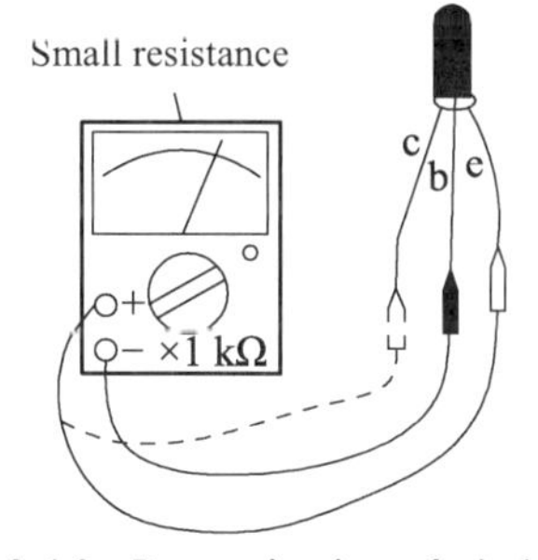

Fig. 2-16 Determination of triode base

In the case of a PNP triode, the red probe shall connect the assumed base b, and the black probe shall connect the other two electrodes. When the resistance measured twice is very small, the red probe is connected to base b, and the triode can be determined to be the PNP triode.

(2) Determine collector c and emitter e.

From the structural diagram of the triode, there is no significant difference between the emitter e and collector c, and they can be interchanged. In fact, there are significant differences in the area and doping concentration between the emitter region and collector region. In case of NPN triode, the black and red probes can be contacted with two undetermined electrodes respectively, and then the black probe and base b can be pinched tightly with fingers (equivalent to connecting a 100 K resistor, be careful not to short-circuit the two electrodes). Observe the swing amplitude of the pointer, as shown in Fig. 2-17(a). Then swap the black and red watch probes and retest them according to the above method. Comparing the swing amplitude of the needle twice. The pin connected to the black probe with a larger swing amplitude is the pole c, while the red probe is connected to pole e. The schematic diagram is as shown in Fig. 2-17(b).

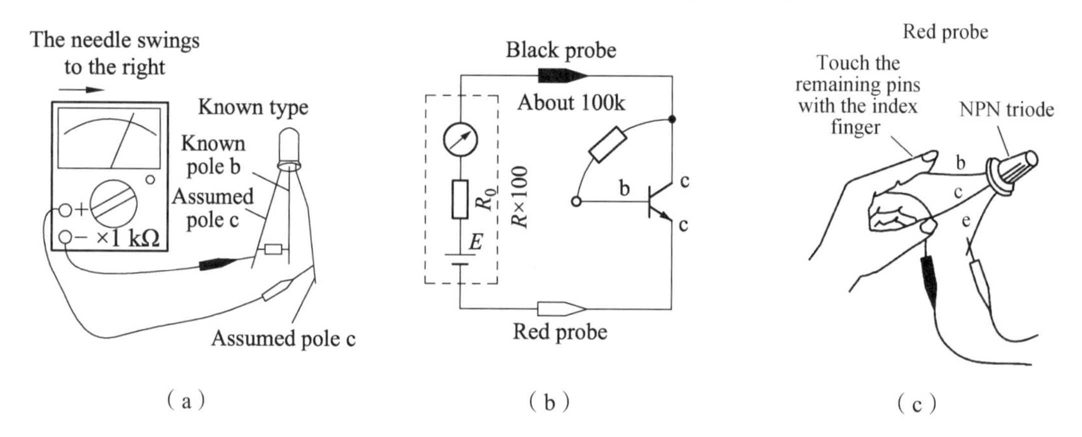

Fig. 2-17　Determination of pole c and pole e

The practice of using fingers to replace the base resistance in actual measurements is as shown in Fig. 2-17(c).

In case of PNP triode, the black and red probes can be swapped in the above method.

2) Quality determination of triode

Using NPN triode as an example to determine the quality of a triode. Use the $R \times 1$ kΩ gear of a multimeter, connect the black probe to the base of the triode, and the red probe to the emitter and collector of the triode. The resistance values measured twice are both at the kΩ level (if the triode is a germanium triode, the resistance value is about 1 kΩ; if it is a silicon triode, the resistance value is about 7 kΩ). Then connect the red probe to the base and the black probe to the pole e and pole c of the triode, respectively. If the measured resistance is infinite, it can be preliminarily determined that the triode is intact. Then use a multimeter to measure the resistance between the pole e and pole c of the triode, and the measured resistance value shall also be infinite. If the measurement results are consistent with the above conclusions, the triode is in good condition.

If the resistance between the triode emitter and the collector is zero or infinite twice, it means that there is a short circuit or open circuit between the triode emitter and the collector, and the triode is no longer available.

In case of PNP triode, simply swap the red and black probes of the multimeter to determine its quality.

5. Precautions for using a multimeter

When using a multimeter, the following points must be noted.

(1) When measuring, the red probe connecting line of the multimeter shall be connected to the red terminal or inserted into the jack marked with " + ". The black probe shall be connected to the black terminal or inserted into the jack marked with " – ".

(2) When using a multimeter, it is generally advisable to hold the probe with hand and avoid touching the metal part of the probe to ensure safety and accurate measurement.

(3) Before use, the changeover switch must be turned to the corresponding gear according to the type and size to be measured.

(4) When measuring current, the instrument shall be connected in series with the circuit; When measuring voltage, the instrument shall be connected in parallel with the circuit. If there is no idea about the item to be measured, turn the changeover switch to the maximum measuring limit for testing, and then select the appropriate measuring limit. Do not conduct the hot-line work when converting the measuring limit to avoid damaging the instrument. When measuring DC current and voltage, pay attention to polarity.

(5) When measuring resistance, live line measurement is not allowed, and the measured resistor cannot have parallel branches. Before measuring the resistance, the ohmmeter shall be zeroed, and after changing the resistance multiplier, it shall be reset to zero.

(6) When AC or pulse voltage is superimposed on the measured DC voltage, attention shall be paid to the withstand voltage of the multimeter to avoid excessive peak voltage damaging the multimeter.

(7) After the use of the multimeter, the changeover switch shall be set to the non-measuring position or the highest gear of AC voltage, so as to avoid damage to the multimeter due to failing to pay attention to the measurement type and measuring limit during the next use.

In short, the multimeter as a common measuring instrument is simple, but it must be used correctly and operated patiently in order to ensure the accuracy of the instrument and improve the accuracy of measurement.

任务二　钳形电流表的检查和使用

一、交流钳形电流表

在电力系统中，根据各专业和工种的不同，人们要从事不同的工作和进行不同的操作，而生产实践又告诉我们，为了顺利完成任务而又不发生人身伤亡事故，操作人员必须携带和使用各种安全用具。

在测量电流时，通常需要将被测电路断开，才能将电流表的一次线圈或电流互感器的一次绕组串接到被测电路中，这在实际的操作中会造成很多不便。而利用钳形电流表则无须断开被测电路就可以测量被测电流，所以在实际中被广泛使用。

交流钳形电流表的外形如图 2-18 所示。它由电流互感器和电流表两部分组成，电流互感器的铁心有一活动部分，与手柄相连。

图 2-18　交流钳形电流表

用交流钳形电流表测量时，只需用手握紧钳形电流表的手柄，电流互感器的铁心便会张开，将被测电流的导线卡入钳口中，然后放开手柄，铁心闭合。此时，被测电流的导线相当于电流互感器的一次绕组，绕在铁心上的二次绕组与电流表相接。电流表所指示的数值取决于二次绕组中电流的大小，而二次绕组电流的大小又与被测电流成正比。所以只要将折算好的刻度作为电流表的刻度，测量时与二次绕组相接的电流表的指针便按比例偏转，指示出被测电流的数值。电流量程可通过量程旋钮 K 选择。

交流钳形电流表的原理电路图如图 2-19 示。图中 TA 为电流互感器的二次绕组。

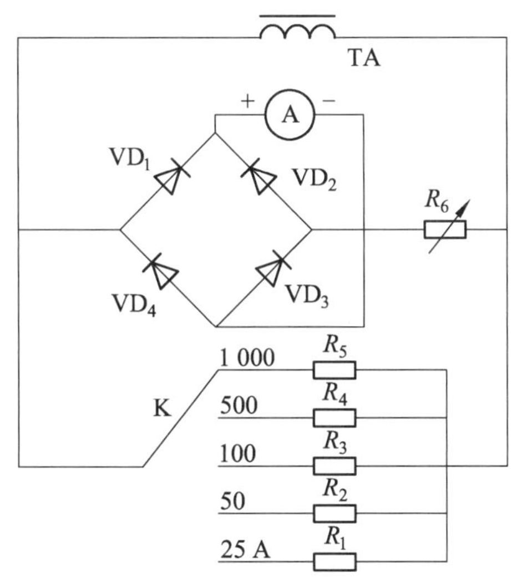

图 2-19　交流钳形电流表原理电路

指示仪表是磁电系电流表；二极管 $VD_1 \sim VD_4$ 构成桥式整流电路，其作用是将电流互感器二次侧的交流电流变换成直流电流，然后由磁电系电流表指示出被测电流的数值；电阻 $R_1 \sim R_5$ 是用于扩大电流量程的分流电阻；R_6 为各电流量程的公共电阻，调节 R_6 可以消除仪表的误差。

由图 2-19 可见，交流钳形电流表是由磁电系电流表表头和桥式整流电路共同构成的整流系电流表。

上述钳形电流表是采用磁电系电流表作为指示仪表，只用于交流电流的测量，如 T301 型钳形电流表。如果采用电磁系电流表作指示仪表，则可以交直流两用。

二、交直流两用钳形电流表

用来测量交直流电流的钳形电流表是由电磁系测量机构构成的，没有二次绕组，如图 2-20 所示。它的外形虽然与交流钳形电流表相同，但结构和工作原理却不一样。在铁心中的被测电流导线相当于电磁系测量机构中的固定线圈，在铁心中产生磁场，其磁通在铁心中形成闭合回路，同时使圆形缺口中间的动铁片磁化。活动铁片与铁心之间的作用，与电磁系排斥型测量机构的作用原理相同，即可动铁片受磁场力的作用发生偏转，从而带动指针指示出被测电流的数值。

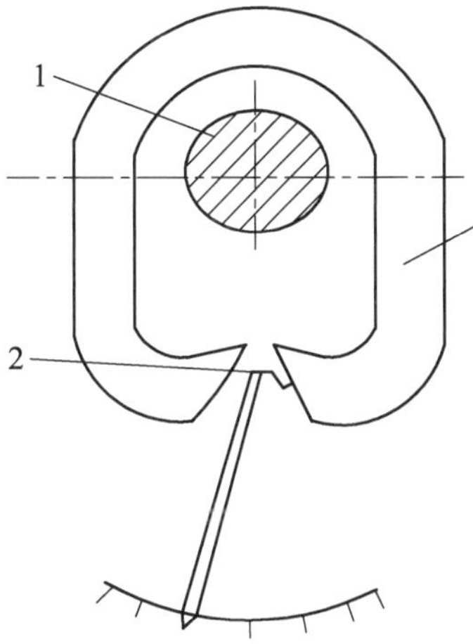

1—被测导线；2—动铁片；3—磁路系统。

图 2-20　交直流两用钳形电流表结构示意图

三、钳形电流表的使用

钳形电流表准确度等级不高，常用于对测量要求不高的场合。在用钳形电流表测量前，应根据被测电流的大小，选择相应的测量量程。假若被测电流大小预先无法估计，则应先将量程旋钮置于大量程进行试测，然后根据被测电流的大小，再变换到合适的量程。但应注意，在整个测量过程中不能随意切换量程，若要变化量程时，应必须先将钳形电流表从被测电路中移去，否则会损坏钳形电流表。

在测量时，被测导线应放在钳口中央，仪器与导线垂直，沿导线方向匀速移动表计观察数值变化（变化量应不大），从而保证数据的准确，以减小误差。钳形电流表盘上标尺刻度通常有多条，读数时要根据量程选择对应的标尺度。所选的量程应能使指针指示在标度尺刻度的 1/2 ~ 2/3，以减小测量时产生的误差。保持固定和活动铁心钳口两个结合面的衔接良好。测量时如果有杂音，则可将钳口重新开合一次；若钳口若有污垢，可用汽油擦净。测量小于 5 A 的电流时，为了获得较准确的测量值，在条件允许的情况下，可将被测导线多绕几圈，再放进钳口进行测量。这时实际的被测电流数值，等于仪表的读数除以放进钳口内的导线根数。不能用钳形电流表测量裸导线中的电流，以防触电和短路。交直流两用钳形电流表要区别使用。

一般不可用钳形电流表测量高压电路中的电流，以免发生事故。在测量时，只能卡一根导线。单相电路中，如果同时卡相线和中性线，则会因两根导线中的电流大小相等、方向相反，使电流表的读数为零。三相对称电路中，同时卡进两相相线，与卡进一相相线时电流读数相同；同时卡进三相相线时读数为零。三相不对称电路中，也只能一相一相地测量，不能同时卡进两相或三相火线。测量完毕，必须把仪表的量程开关置于最大量程位置上，以防下次使用时，因疏忽大意未选择量程就进行测量，而造成损坏仪表的意外事故。

四、多用途钳形表

（一）交流电流、电压钳形表

钳形表不仅可以测量交流电流，还可以测量交流电压，如 T-302、MG-24 都是交流电流、电压钳形表。多用途钳形表的侧面有电压测量的接线插口。测量交流电压时，不用电流互感器，改用两根钳形表表笔插入电压插口接线测量。

如图 2-21 所示，是测量交流电流、电压的多用途钳形表工作原理图。图中电阻 $R_1 \sim R_5$ 是测量电流时的各量程的分流电阻，$R_8 \sim R_{10}$ 是电压表各电压量程的附加电阻。

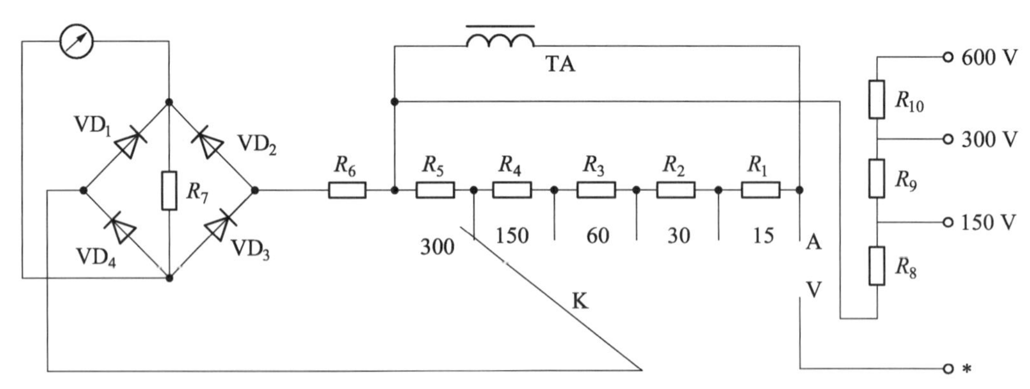

图 2-21　交流电流、电压钳形表工作原理

（二）电流、电压、功率三用钳形表

这种表不但能测量电流、电压，还能测量功率，如 MG4-1 型三用钳形表。JQD-85A 型除可测量电流、电压外，还可测量功率因数和功率。

（三）多用钳形表

多用钳形表由钳形电流互感器和万用表组合而成。将互感器拔除，将表笔棒插入仪表上方插孔内，便可作为万用表使用。如 MG-28 型可以测量交直流电流、交直流电压和电阻。

多用钳形表使用方便，不用切断电路，就能测出电路或设备中的电流、电压及其他电量，但在测量过程中不能换挡，并且只能对电气设备或电路的运行情况做粗略测量。

五、常用钳形电表的技术特性

表 2-4 是几种常用钳形电表的主要技术特性。

表 2-4　常用钳形电表技术特性

名称	型号	准确度等级	测量范围	1 min 绝缘耐压/V
钳形交流电流表	T-301	2.5	0～10～25～50～100～250 A 0～10～25～100～300～600 A 0～10～30～100～300～1 000 A	2 000
钳形交直流电流、电压表	T-302	2.5	电流：0～10～30～100～300～1 000 A 电压：0～250～500 V 0～300～600 V	2 000
袖珍型钳形表	MG-24	2.5	电流：0～5～25～250 A 电压：0～300～600 V	2 000
袖珍型三用钳形表	MG-25	2.5	交流电流：5～25～100 A 交流电压：300～600 V	2 000

Task Ⅱ Inspection and use of clip-on ammeter

Ⅰ. AC clip-on ammeter

In the electric power system, according to the different disciplines and types of work, people have to engage in different work and different operations, and the production practice tells us that in order to successfully complete the task without personal injury or death, the operators must carry and use a variety of safety appliances.

When measuring the current, it is usually necessary to disconnect the circuit under test to connect the primary coil of the ammeter or the primary winding of the current transformer to the circuit under test, which causes a lot of inconvenience in the actual operation. The clip-on ammeter can measure the current under test without disconnecting the circuit under test, so it is widely used in practice.

The outline drawing of AC clip-on ammeter is as shown in Fig. 2-18. It consists of a current transformer and an ammeter. The iron core of the current transformer has a moving part, which is connected with the handle.

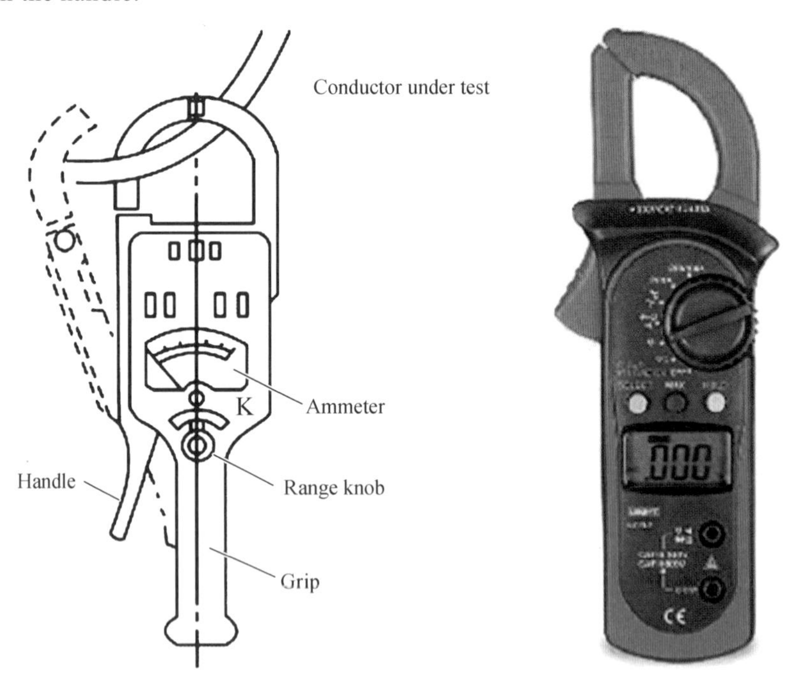

Fig. 2-18 AC clip-on ammeter

When measuring with an AC clip-on ammeter, just hold the handle of the clip-on ammeter with hand, the iron core of the current transformer will be opened. Clip the conductor of the current under test into the clip, then release the handle and the iron core closes. At this time, the conductor of the current under test is equivalent to the primary winding of the current transformer, and the secondary winding wound on the iron core is connected with the ammeter. The value indicated by the ammeter depends on the current in the secondary winding, and the current in the secondary winding is proportional to the measured current. So long as the converted scale is taken as the scale of the ammeter, the pointer of the ammeter connected to the secondary winding will deflect proportionally to indicate the value of the measured current. The current range can be selected by the range knob K.

The schematic circuit diagram of the AC clip-on ammeter is as shown in Fig. 2-19. TA in the figure represents the secondary winding of the current transformer.

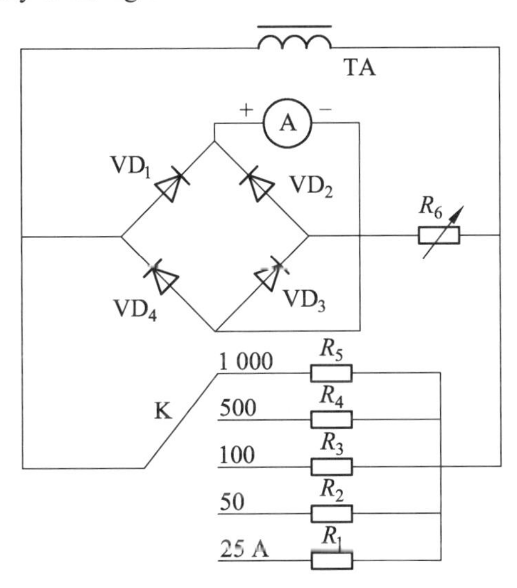

Fig. 2-19 Schematic circuit of AC clip-on ammeter

The indicating instrument is a magnetoelectric ammeter; The diodes $VD_1 \sim VD_4$ form a bridge rectifier circuit, whose function is to convert the AC current on the secondary side of the current transformer into DC current, and then the value of the measured current is indicated by the magnetoelectric ammeter. Resistors $R_1 \sim R_5$ are shunt resistors used to expand the current range; R_6 is the common resistor of each current range. Adjusting R_6 can eliminate the error of the instrument.

It can be seen from Fig. 2-19 that the AC clip-on ammeter is a rectification ammeter composed of magnetoelectric ammeter head and bridge rectifier circuit.

The above clamp Ammeter adopts magnetoelectric ammeter as the indicating instrument, which is only used for AC current measurement, such as T301 clip-on ammeter. If the electromagnetic ammeter is used as the indicating instrument, it can be used for both AC and DC.

Ⅱ. DC-AC clip-on ammeter

The clip-on ammeter used to measure AC and DC current is composed of electromagnetic measuring mechanism and has no secondary winding, as shown in Fig. 2-20. Although its outline is the same as that of the AC clip-on ammeter, its structure and working principle are different. The measured current conductor in the iron core is equivalent to a fixed coil in an electromagnetic measuring mechanism, generating a magnetic field in the iron core. Its magnetic flux forms a closed circuit in the iron core, while magnetizing the moving iron sheet in the middle of the circular notch. The action between the moving iron sheet and the iron core is the same as that of the electromagnetic repulsive measuring mechanism, that is, the moving iron sheet is deflected under the action of the magnetic field force, thus driving the pointer to indicate the value of the measured current.

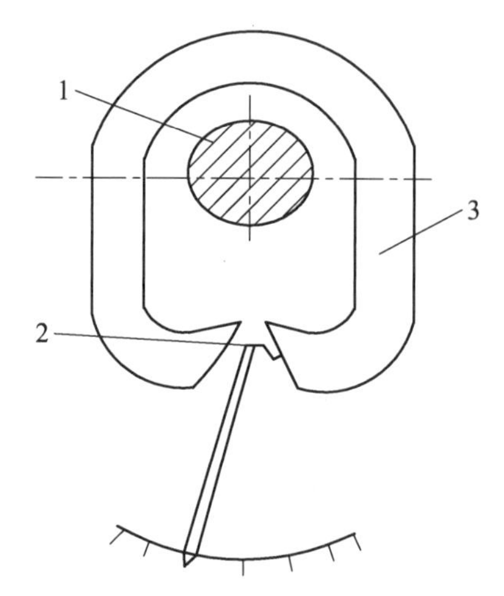

1—conductor under test; 2—moving iron sheet; 3—magnetic circuit system.

Fig. 2-20 Schematic diagram of structure of DC-AC clip-on ammeter

Ⅲ. Use of clip-on ammeter

The accuracy level of the clip-on ammeter is not high, so it is often used in situations where the measurement requirement is not high. Before measuring with a clip-on ammeter, the corresponding measuring range shall be selected according to the magnitude of the current being measured. If the magnitude of the measured current cannot be estimated in advance, the range knob shall be placed on a large range for testing, and then changed to a suitable range based on the magnitude of the measured current. However, it shall be noted that the range can not be switched at will during the whole measurement process. To change the range, the clip-on ammeter must be removed from the tested circuit, otherwise the clip-on ammeter will be damaged.

During measurement, the conductor under test shall be placed in the center of the clip, with the instrument perpendicular to the conductor. The ammeter shall be moved at a uniform speed along the direction of the conductor to observe that there is little change in numerical values, ensuring the accuracy of the data and reducing errors. There are usually multiple scales on the clip-on ammeter, and the corresponding scale shall be selected according to the range when reading. The selected range shall enable the pointer to indicate more than 1/2–2/3 of the scale to reduce the measurement error. Maintain a good connection between the two joints of fixed and movable iron core clips. If there is a noise during the measurement, the clip can be opened and closed again. If there is dirt on the clip, wipe it off with gasoline. When measuring the current less than 5 A, in order to obtain a more accurate measured value, if the conditions permit, the conductor under test can be wound several more times and then put into the clip for measurement. At this time, the actual measured current value is equal to the instrument reading divided by the number of conductors placed in the clip. The current in the bare conductor cannot be measured with a clip-on ammeter to prevent electric shock and short circuit. AC-DC clip-on ammeter shall be used separately.

Generally speaking, the clip-on ammeter can not be used to measure the current in the HV circuit to avoid accidents. During measurement, only one conductor can be clamped. In a single-phase circuit, if the phase line and neutral wire are clamped in at the same time, the reading of the ammeter will be zero because the current in the two conductors is equal and reversed. In a three-phase symmetrical circuit, the current reading is the same when the two phase lines are clamped in at the same time as when the one phase line is clamped in; The reading is zero when three phase lines are clamped in. In the three-phase asymmetric circuit, it can only be measured phase by phase, and two phase or three phase live wires can not be clamped in at the same time. After the measurement, the range switch of the instrument must be placed at the maximum range position, so as to prevent the accidental damage to the instrument caused by carelessness in the measurement without selecting the range next time.

Ⅳ. Multi-purpose clip-on meter

(Ⅰ) AC clip-on voltammeter

Clip-on voltammeter can measure not only AC current, but also AC voltage. For example, T-302 and MG-24 are AC clip-on voltammeters. The side of the multi-purpose clip-on meter has a wiring jack for voltage measurement. When measuring AC voltage, instead of using a current transformer, insert two clip-on meter probes into the voltage jack for wiring measurement.

As shown in Fig. 2-21, it is a working principle schematic diagram of a multi-purpose clip-on meter for measuring AC current and voltage. The resistors $R_1 \sim R_5$ in the figure are the shunt resistors of each range when measuring the current, and $R_8 \sim R_{10}$ are the additional resistors of each voltage range of the voltmeter.

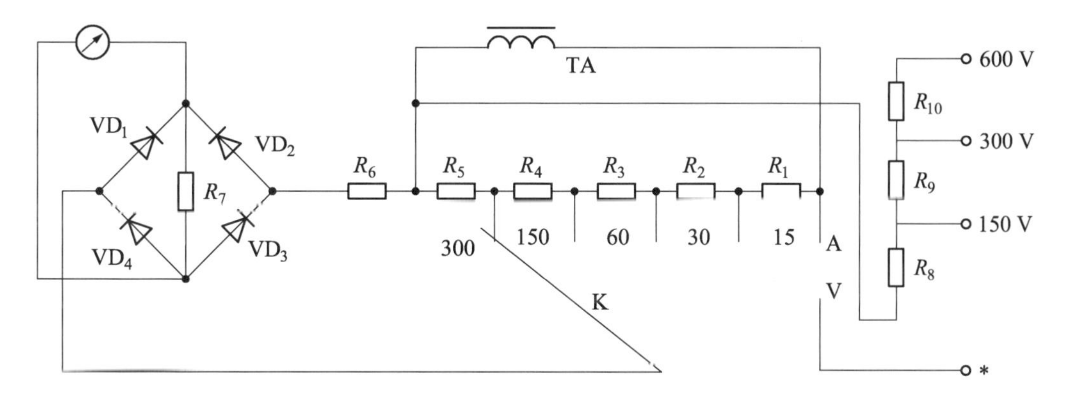

Fig. 2-21 Working principle of AC clip-on voltammeter

(Ⅱ) Current-voltage-power three-purpose clip-on meter

This meter can measure not only current and voltage, but also power, such as MG4-1 three-purpose clip-on meter. JQD-85A can measure not only current and voltage, but also power factor and power.

(Ⅲ) Multi-purpose clip-on meter

The multi-purpose clip-on meter is composed of a clip-on current transformer and a multimeter. Remove the mutual inductor and insert the probe into the jack above the instrument, which can be used as a multimeter. MG-28 can measure AC/DC current, AC/DC voltage, and resistance.

The multi-purpose clip-on meter is easy to use. It can measure the current, voltage, and other electric quantities in the circuit or equipment without cutting off the circuit, but it can not change gears in the measurement process, and can only roughly measure the operation of the electrical equipment or circuit.

Ⅴ. Technical characteristics of common clip-on meters

Tab. 2-4 shows the main technical characteristics of several common clip-on meters.

Tab. 2-4 Technical characteristics of common clip-on meters

Name	Model	Accuracy level	Measurement range	1 min insulation withstand voltage/V
AC clip-on ammeter	T-301	2.5	0-10-25-50-100-250 A 0-10-25-100-300-600 A 0-10-30-100-300-1,000 A	2,000
AC-DC clip-on voltammeter	T-302	2.5	Current: 0-10-30-100-300-1,000 A Voltage: 0-250-500 V; 0-300-600 V	2,000
Pocket-size clip-on meter	MG-24	2.5	Current: 0-5-25-250 A Voltage: 0-300-600 V	2,000
Pocket-size three-purpose clip-on meter	MG-25	2.5	AC current: 5-25-100 A AC voltage: 300-600 V	2,000

任务三　绝缘电阻表的检查和使用

绝缘电阻表适用于测量大电阻和绝缘电阻，计量单位是兆欧，用“MΩ”表示。电气设备绝缘性能的好坏，关系到设备正常运行和工作人员的人身安全。绝缘电阻的阻值都比较大，从几十到几千兆欧，在这个范围内万用表欧姆挡的刻度很不准确，况且万用表测量电阻所用的电源电压比较低，在低电压下呈现的绝缘电阻值不能反映在高电压作用下的绝缘电阻的真实值，因此测量绝缘电阻不能用万用表，而必须用具有高压电源的绝缘电阻表。各种电压等级的电气设备和线路的绝缘电阻的大小都有具体的规定，一般来说，绝缘电阻越大，绝缘性能越好。常用的绝缘电阻表的外形如图 2-22（a）所示，测量接线方法如图 2-22（b）（c）（d）所示。

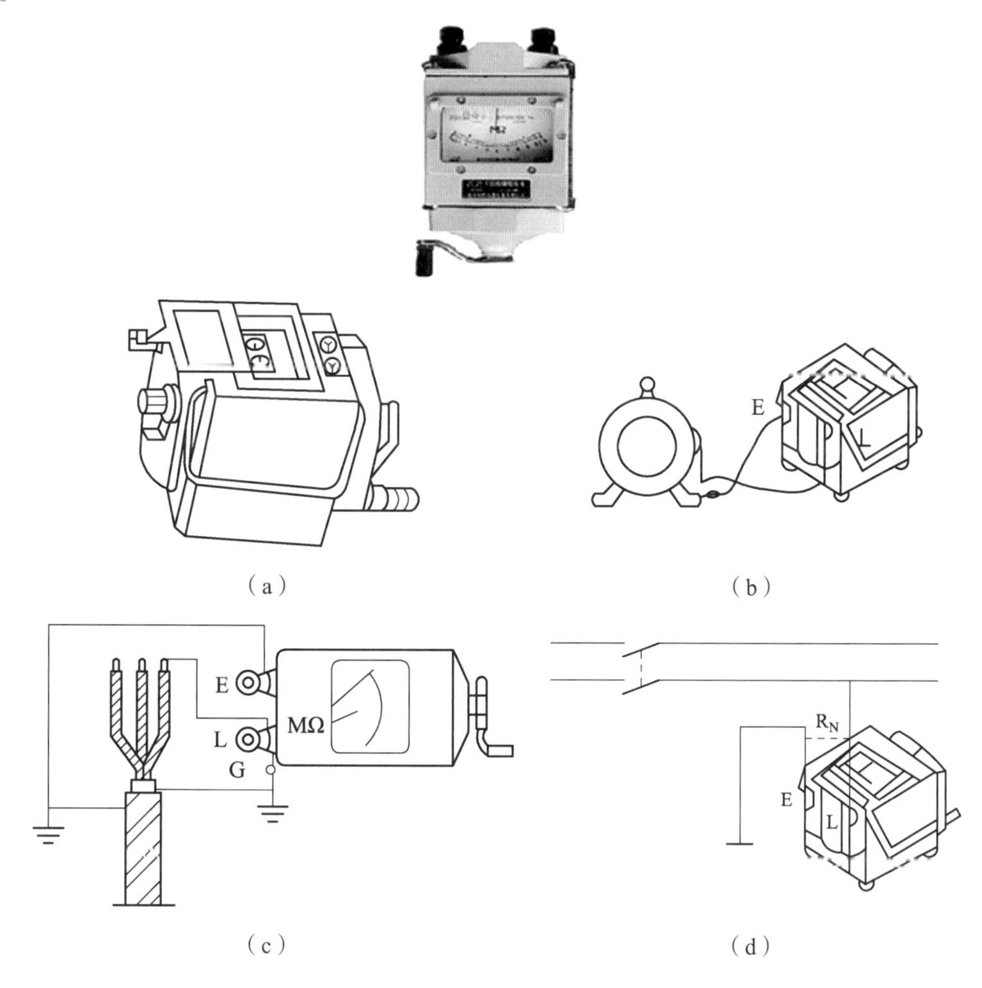

图 2-22　绝缘电阻表及绝缘电阻表的测量接线方法

一、绝缘电阻表的结构

绝缘电阻表的种类很多，但其结构基本相同，都是由比率型磁电系测量机构和手摇发电机两个部分组成。

比率型磁电系测量机构是一种特殊形式的磁电系测量机构，如图 2-23 所示为比率型磁电系测量机构的结构示意图。固定部分由永久磁铁、极掌和铁心等部件组成，由于铁心开口，极掌和铁心的形状比较特殊，故铁心与磁极间的气隙中的磁场不均匀。可动部分有两个可动线圈，它们彼此相交成一固定角度，并连同指针一起固定装在同一转轴上。当两个线圈通有电流时，线圈 1 中的电流 I_1 与气隙磁场相互作用产生转动力矩 M_1，线圈 2 中的电流 I_2 与气隙磁场相互作用产生反作用力矩 M_2。在这两个力矩的共同作用下，可动部分发生偏转，指针也随之偏转；当两个力矩平衡时，指针就停留在一个稳定的位置上。

由于气隙中磁场分布不均匀，所以转动力矩 M_1 不仅与电流 I_1 成正比，还与线圈 1 所处的位置有关，即 M_1 随可动部分的偏转角度 α 不同而变化，所以得

$$M_1 = I_1 f_1(\alpha) \tag{2-5}$$

同理 $M_2 = I_2 f_2(\alpha)$

当两个力矩平衡时，有：

$$I_1 f_1(\alpha) = I_2 f_2(\alpha)$$

$$\frac{I_1}{I_2} = \frac{f_1(\alpha)}{f_2(\alpha)} = f(\alpha)$$

$$\alpha = f\left(\frac{I_1}{I_2}\right) \tag{2-6}$$

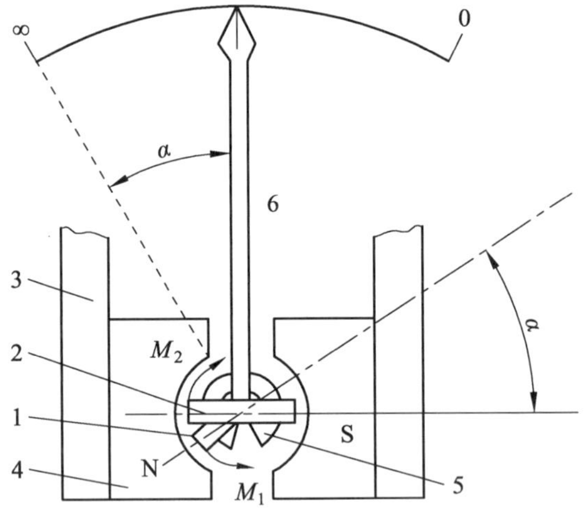

1、2—动圈；3—永久磁铁；4—极掌；5—带缺口的圆柱形铁心；6—指针。

图 2-23　比率型磁电系测量机构的结构示意图

由式（2-6）可知，磁电系测量机构的可动部分的偏转角度及只决定于两个动圈电流 I_1 与 I_2 的比值，而与其他因素无关。即使因发电机的手摇速度不稳定而造成输出电压波动，由于 I_1 与 I_2 同时变化，对仪表的指示值也不会有影响。由于这种测量机构没有机械游丝，所以它不通电时，其指针可停留在任意位置。这种仪表的指示值取决于两线圈电流的比值，所以又被称为比率表或流比计。

绝缘电阻表的手摇发电机一般为直流发电机或交流发电机和整流电路结合的装置，其容量很小，但输出电压却很高，绝缘电阻表以发电机的额定电压来分类，常见的有 500 V、1 000 V、2 500 V、5 000 V 几种。一般发电机都设有离心调速装置，以保证转子能在额定转速下恒速转动。

二、绝缘电阻表的测量原理

绝缘电阻表的测量原理电路图如图 2-24 所示。

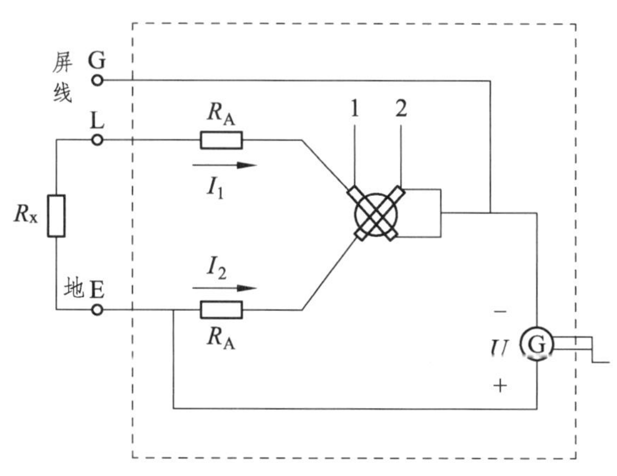

图 2-24　绝缘电阻表的测量原理电路图

绝缘电阻表的内部有两个回路。一个回路从正电源端经被测电阻 R_x、限流电阻 R_A、动圈 1 回到负电源端；另一回路从正电源端经附加电阻 R_v、动圈 2 回负电源端。当手摇发电机输出 定的直流电压时，在两个动圈中产生的电流分别为

$$I_1 = \frac{U}{R_x + R_A + r_1}$$

$$I_2 = \frac{U}{R_v + r_2} \qquad (2\text{-}7)$$

式中，r_1、r_2 分别为动圈 1 和 2 的内阻。

只要发电机的输出电压 U 大小不变，则 I_1 将随被测电阻 R_x 的增大而减小；而 I_2 却是一个与被测电阻 R_x 无关的常量。将 I_1 和 I_2 代入 $\alpha = f\left(\frac{I_1}{I_2}\right)$ 得

$$\alpha = f\left(\frac{I_1}{I_2}\right) = f\left(\frac{R_2 + r_2}{R_x + R_A + r_1}\right) = f(R_x) \qquad (2\text{-}8)$$

由于 r_1、r_2、R_v 和 R_x 均为常数，所以可动部分的偏转角 α 只与被测电阻 R_x 有关，即偏转角 α 能直接反映绝缘电阻的大小。

当被测电阻 R_x=0 时，相当于 L（线）与 E（地）两端子短接。此时，电流 I_1 最大，可动部分的偏转角 α 也最大，指针偏转到标度尺的最右端。

当被测电阻 R_x=0 时，相当于 L（线）与 E（地）两端子开路。此时，电流 I_1=0，可动部分在 I_2 的作用下，指针偏转到标度尺的最左端。由此可见，绝缘电阻表的标度尺是反向刻度的。

从上述可知，绝缘电阻表的标度尺的刻度是不均匀的。从理论上来说，绝缘电阻表的测

量范围为 0 ~ ∞，但实际上只有部分刻度较为准确，因此，在技术要求中一般都标明绝缘电阻表的准确度范围。

三、绝缘电阻表的使用

（一）绝缘电阻表的选用

绝缘电阻表按其额定电压分 500 V、1 000 V、2 500 V、5 000 V 几种。通常应根据被试设备的额定电压来选择绝缘电阻表，绝缘电阻表的额定电压过高，可能在测试中损坏被试设备的绝缘。一般来说，测量额定电压在 500 V 及以下的设备的绝缘电阻时，可选用 500V 绝缘电阻表；测量额定电压在 500 V 至 1 000 V 的设备的绝缘电阻时，可选用 1 000 V 绝缘电阻表；测量额定电压在 1 000 V 以上设备的绝缘电阻时，可选用 2 500 V 的绝缘电阻表；测量额定电压在 35 kV 及以上的设备的绝缘电阻时，可选用 5 000 V 的绝缘电阻表。

（二）绝缘电阻表的接线和测量方法

绝缘电阻表有 3 个接线柱，其中两个较大的接线柱上分别标有接地（E）和线路（L），另一个较小的接线柱上标有保护环或屏蔽（G）。

（1）测量电机的绝缘电阻。将绝缘电阻表的（E）接线柱接机壳，（L）接线柱接到电机绕组上，如图 2-22（b）所示。线路接好后，可按顺时针方向摇动绝缘电阻表的发电机手柄，转速由慢到快，直到 120 r/min 的均匀速度，当绝缘电阻表的发电机转速稳定时，表针也稳定下来，这时表针指示的数值就是所测得的绝缘电阻值。

（2）测量电缆的绝缘电阻。测量电缆的导电线芯与电缆外壳的绝缘电阻时，除将被测两端分别接 E 和 L 两接线柱外，还需将 G 接线柱引线接到电缆壳与芯之间的屏蔽层上，E 和 G 接线柱应可靠接地，如图 2-22（c）所示。

（3）测量线路绝缘电阻。如图 2-22（d）所示。

（三）使用绝缘电阻表的注意事项

（1）测量电气设备和线路的绝缘电阻时，必须先切断电源，然后将设备进行放电，以保证人身安全和测量准确。

（2）绝缘电阻表使用时应放在水平位置，未接线前先转动绝缘电阻表做开路试验，指针应指在∞处，再将 L 和 E 两个接线柱短接，慢慢转动绝缘电阻表，看指针是否指在 0 处，若能指在 0 处，说明绝缘电阻表是好的。

（3）绝缘电阻表接线柱上引出线应用多股软线，且要有良好的绝缘，两根引线切忌绞在一起，避免造成测量数据的不准确。

（4）绝缘电阻表测量完后应先断开接至被试品高压端的连接线，然后将绝缘电阻表停止转动，随后对被测物放电，在绝缘电阻表的摇把没有停止转动和被测物没有放电前，不可用手拆除引线或触及被测物的测量部分，以防触电。

Task Ⅲ Inspection and use of insulation resistance meter

The insulation resistance meter is suitable for measuring large resistance and insulation resistance, and the unit of measurement is megohm, which is represented by "MΩ". The insulation of electrical equipment is related to the normal operation of the equipment and the personal safety of the workers. The values of insulation resistance are relatively large, ranging from tens to thousands of megohms, in this range, the ohmic gear of the multimeter is very inaccurate. Moreover, the power supply voltage used for measuring resistance with a multimeter is relatively low, and the insulation resistance value presented at low voltage cannot reflect the true value of insulation resistance under high voltage. Therefore, a multimeter cannot be used for measuring insulation resistance, and an insulation resistance meter with a high-voltage power supply must be used instead. There are specific regulations on the insulation resistance of electrical equipment and circuits of all voltage classes. Generally speaking, the greater the insulation resistance, the better the insulation. The outline of the common insulation resistance meter is as shown in Fig. 2-22(a), and the method of measuring wiring is as shown in Fig. 2-22(b)(c)(d).

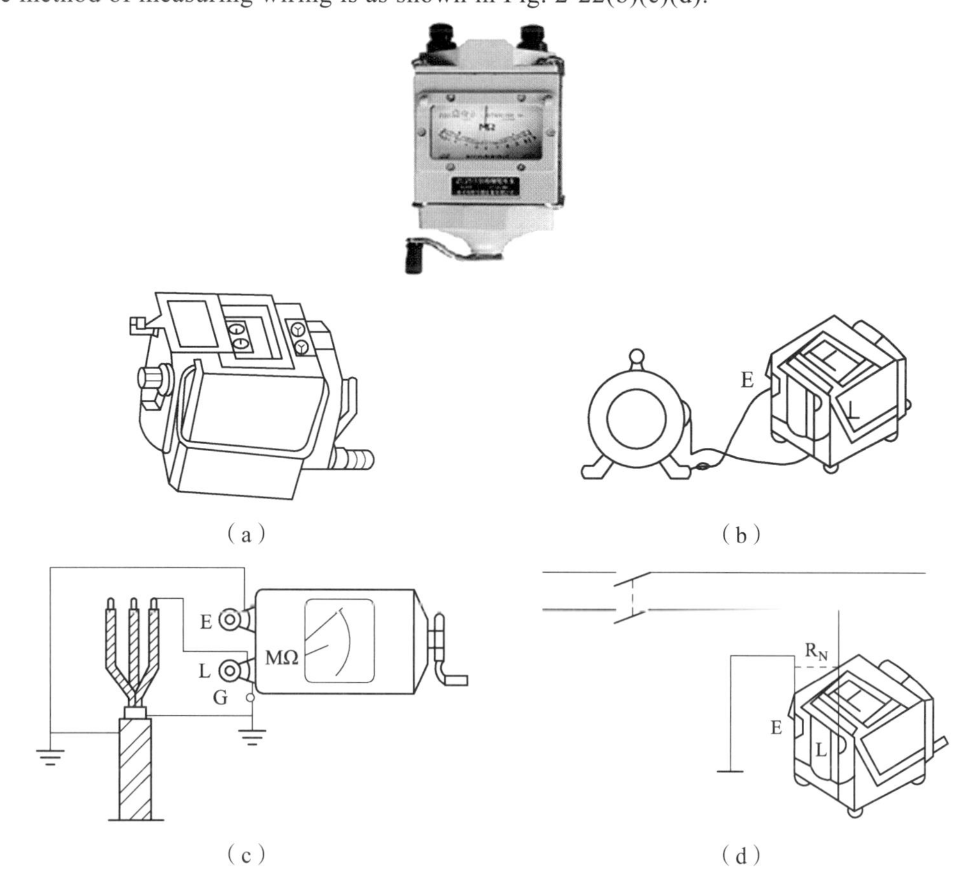

Fig. 2-22 Insulation resistance meter and method of measuring wiring of insulation resistance meter

Ⅰ. Structure of insulation resistance meter

There are many types of insulation resistance meters, but their structures are basically the same, consisting of two parts: a ratio type magnetoelectric measuring mechanism and a hand generator.

The ratio type magnetoelectric measuring mechanism is a special form of magnetoelectric measuring mechanism. Fig. 2-23 is a schematic diagram of the structure of a ratio type magnetoelectric measuring mechanism. The fixed part consists of permanent magnet, pole shoe, iron core, and other components. Due to the opening of the iron core, the shape of the pole shoe and iron core is relatively special, so the magnetic field in the air gap between the iron core and the magnetic pole is uneven. The movable part has two movable coils, which intersect each other at a fixed angle and are fixed on the same pivot together with the pointer. When there is a current in both coils, the current I_1 in coil 1 interacts with the air-gap magnetic field to produce a rotating torque, and the current I_2 in coil 2 interacts with the air-gap magnetic field to produce a reaction torque M_2. Under the combined action of these two torques, the movable part deflects, and the pointer also deflects accordingly. When the two torques are balanced, the pointer stays in a stable position.

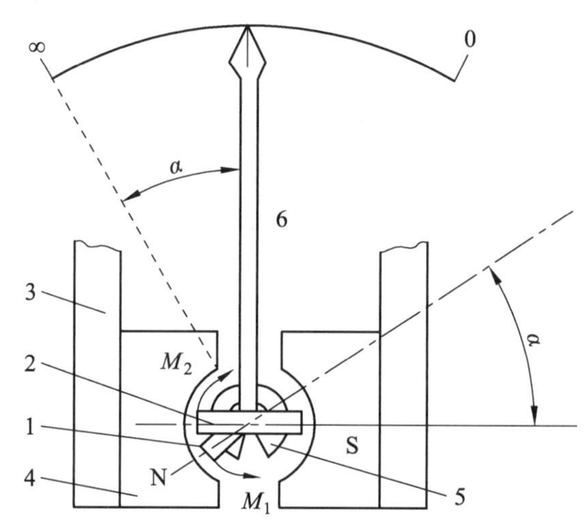

1, 2—movable coils; 3—permanent magnet; 4—pole shoe; 5—cylindrical iron core with notch; 6—pointer.

Fig. 2-23 Schematic diagram of the structure of a ratio type magnetoelectric measuring mechanism

Due to the uneven distribution of magnetic field in the air gap, the rotating torque M_1 is not only proportional to the current I_1, but also related to the position of the coil 1, that is to say, M_1 varies with the deflection angle a of the movable part, so:

$$M_1 = I_1 f_1(\alpha) \tag{2-5}$$

Similarly, $$M_2 = I_2 f_2(\alpha)$$

When two torques are balanced, there are:

$$I_1 f_1(\alpha) = I_2 f_2(\alpha)$$

$$\frac{I_1}{I_2}=\frac{f_1(\alpha)}{f_2(\alpha)}=f(\alpha)$$

$$\alpha=f\left(\frac{I_1}{I_2}\right) \quad (2\text{-}6)$$

From the above equation, it can be seen that the deflection angle of the movable part of the magnetoelectric measuring mechanism is only determined by the ratio of the two movable coil currents I_1 and I_2, and is independent of other factors. Even if the output voltage fluctuates due to the unstable hand speed of the generator, the indicator value of the instrument will not be affected because I_1 and I_2 change at the same time. Because this measuring mechanism has no mechanical hairspring, when it is not electrified, its pointer can stay at any position. The indicated values of this instrument depend on the ratio of the current between the two coils, so it is also called a ratio meter or quotient meter.

The hand generator of insulation resistance meter is usually a combination of DC generator or alternator and rectifier circuit, which has a small capacity but a high output voltage. Insulation resistance meters are classified based on the rated voltage of the generator, commonly including 500 V, 1,000 V, 2,500 V, and 5,000 V. In general, generators are equipped with centrifugal speed governing devices to ensure that the rotor can rotate at a constant speed at the rated speed.

Ⅱ. Measurement principle of insulation resistance meter

The measuring principle circuit diagram of the insulation resistance meter is as shown in Fig. 2-24.

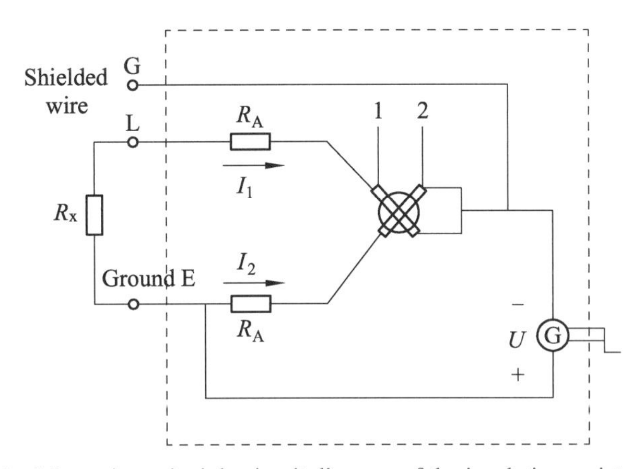

Fig. 2-24 Measuring principle circuit diagram of the insulation resistance meter

There are two circuits inside the insulation resistance meter. A circuit returns to the negative power supply terminal through the measured resistor R_x, current limiting resistor R_A, and movable coil 1 from the positive power supply terminal; The other circuit returns to the negative power supply terminal through the additional resistor R_v and movable coil 2 from the positive power supply terminal. When the hand generator outputs a certain DC voltage U, the currents generated in the two movable coils are:

$$I_1 = \frac{U}{R_x + R_A + r_1}$$

$$I_2 = \frac{U}{R_v + r_2} \qquad (2\text{-}7)$$

Where, r_1 and r_2 are the internal resistances of movable coils 1 and 2, respectively.

As long as the output voltage U of the generator remains constant, I_1 will decrease with the increase of the measured resistance R_x; However, I_2 is a constant independent of the measured resistance R_x. Substituting I_1 and I_2 into $\alpha = f\left(\frac{I_1}{I_2}\right)$ yields:

$$\alpha = f\left(\frac{I_1}{I_2}\right) = f\left(\frac{R_2 + r_2}{R_x + R_A + r_1}\right) = f(R_x) \qquad (2\text{-}8)$$

Because r_1, r_2, R_v and R_A are all constant, the deflection angle a of the movable part is only related to the measured resistance R_x, that is, the deflection angle a can directly reflect the insulation resistance.

When the measured resistance R_x=0, it is equivalent to a short circuit between the L (line) and E (ground) terminals. At this time, the current I_1 is the largest, the deflection angle a of the movable part is also the largest, and the pointer is deflected to the rightmost end of the scale.

When the measured resistance $R_x=\infty$, it is equivalent to an open circuit between the L (line) and E (ground) terminals. At this time, the current I_1=0, and under the action of I_2, the movable part deflects the pointer to the leftmost end of the scale. Thus it can be seen that the scale of the insulation resistance meter is in reverse.

From the above, it can be seen that the scale of the insulation resistance meter is uneven. Theoretically, the measuring range of insulation resistance meter is $0-\infty$, but in practice, only part of the scale is more accurate. Therefore, the accuracy range of insulation resistance meter is generally indicated in the technical requirements.

Ⅲ. Use of insulation resistance meter

(I) Selection and use of insulation resistance meter

Usually, insulation resistance meters are divided into several types based on their rated voltage: 500 V, 1,000 V, 2,500 V, and 5,000 V. Usually, insulation resistance meters shall be selected based on the rated voltage of the equipment under test. If the rated voltage of the insulation resistance meter is too high, it may damage the insulation of the equipment under test during test. Generally speaking, when measuring the insulation resistance of equipment with a rated voltage of 500 V or less, 500 V insulation resistance meter can be selected. When measuring the insulation resistance of equipment with rated voltages from 500 V to 1,000 V, 1,000 V insulation resistance meter can be selected. When measuring the insulation resistance of equipment with a rated voltage above 1,000 V, 2,500 V insulation resistance meter can be selected. When measuring the insulation resistance of equipment with a rated voltage of 35 kV and above, 5,000 V insulation resistance meter can be selected.

(Ⅱ) Wiring and measurement methods for insulation resistance meter

An insulation resistance meter has three terminals, of which two larger terminals are marked with grounding (E) and line (L), respectively, and the other smaller one is marked with a protective ring or shield (G).

(1) Measure the insulation resistance of the motor. Connect the (E) terminal of the insulation resistance meter to the casing, and the (L) terminal to the motor winding, as shown in Fig. 2-22(b). After the circuit is connected, the generator handle of the insulation resistance meter can be shaken clockwise, with the speed increasing from slow to fast until a uniform speed of 120r/min is achieved. When the generator speed of the insulation resistance meter is stable, the needle is also stable, and the value indicated by the needle is the measured insulation resistance.

(2) Measure the insulation resistance of the cable. When measuring the insulation resistance between the conductive core of the cable and the cable jacket, in addition to connecting the tested two ends to the E and L terminals, it is also necessary to connect the G terminal lead to the shielding layer between the cable jacket and the core. The E and G terminals shall be reliably grounded, as shown in Fig. 2-22(c).

(3) Measure line insulation resistance, as shown in Fig. 2-22(d).

(Ⅲ) Precautions for using insulation resistance meter

(1) When measuring the insulation resistance of electrical equipment and lines, it is necessary to cut off the power supply first and then discharge the equipment to ensure personal safety and accurate measurement.

(2) The insulation resistance meter shall be placed in a horizontal position when in use, and the insulation resistance meter shall be turned to conduct the open circuit test before being connected, and the pointer shall be pointed at ∞, then the L and E terminals shall be connected short, and the insulation resistance meter shall be rotated slowly to see if the pointer is pointed at 0. If it can be pointed at 0, it means that the insulation resistance meter is in good condition.

(3) The lead wires on the insulation resistance meter terminal shall be multi-stranded flexible wires with good insulation. The two leads shall not be twisted together to avoid inaccurate measured data.

(4) After measurement with the insulation resistance meter, the connecting wire to the high-voltage end of the test object shall be disconnected first, and then the insulation resistance meter shall stop rotating. Then, the measured object shall discharge. Before the rocker of the insulation resistance meter does not stop rotating and the measured object does not discharge, the lead shall not be removed by hand or the measuring part of the measured object shall not be touched to prevent electric shock.

任务四　接地电阻表的检查和使用

电力系统中的接地按作用不同一般分为 3 种，即工作接地、保护接地和防雷接地。为了保证电气设备可靠运行而将系统中某一点接地称为工作接地。电气设备在运行中，因各种原因其绝缘可能发生击穿和漏电而使设备外壳带电，危及人身和设备安全，因此一般都要求将电气设备的外壳接地，这种接地称为保护接地。为了防止雷电袭击，在电气设备或输电线路上都装有避雷装置，而这些避雷装置也要可靠接地，这种接地称为防雷接地。在上述接地系统中，接地电阻的大小直接关系到人身和设备的安全，其大小与大地的结构、土壤的电阻率、接地体的几何尺寸及形状等因素有关，各种不同电压等级的电气设备和输电线路对接地电阻的标准要求在规程中都有相应的规定，如果接地电阻不符合要求，不仅不能保证安全，还会造成安全错觉，形成事故隐患。因此，必须定期测量接地电阻。

一、接地及接地电阻的概念

所谓接地就是用金属导线将电气设备和输电线路需要接地的部分与埋在土壤中的金属接地体连接起来。接地体的接地电阻包括接地体本身电阻、接地线电阻、接地体与土壤的接触电阻和大地的散流电阻。由于前三项电阻很小，可以忽略不计，故接地电阻一般就指散流电阻。

当接地体上有电压时，就有电流从接地体流入大地并向四周扩散，如图 2-25 所示。越靠近接地体，电流通过的截面越小，电阻越大，电流密度就越大，地面电位也越高；离开接地体越远，电流通过的截面越大，电阻越小，电流密度就越小，电位也越低。到离开接地体大约 20m 处，电流密度几乎等于零，电位也就接近于零，所以接地电阻主要就是从接地体到零电位点之间的电阻。它等于接地体的对地电压与经接地体流入大地中的接地电流之比 $\left(R=\dfrac{U}{I}\right)$。对地电压就是电气设备的接地点与大地零电位之间的电位差。

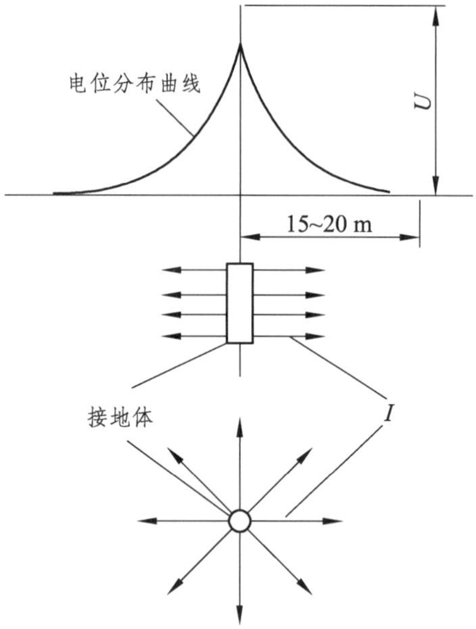

图 2-25　接地电流和电位分布

接地电阻测量线路原理如图 2-26 所示，在被测接地体 E 几十米以外的地方向地下插入辅助接地极 C，并将交流电压加于 E、C 端，于是将有电流 I 通过电极和大地，从接地体 E 出来的电流路线分散在各个不同的方向，离开 E 极越远，电流密度越小。由于在距接地体 E 越远的地方电阻越小，而距接地体 E 越近的地方电阻就越大，所以电压大部分降落在接地体附近的地带。

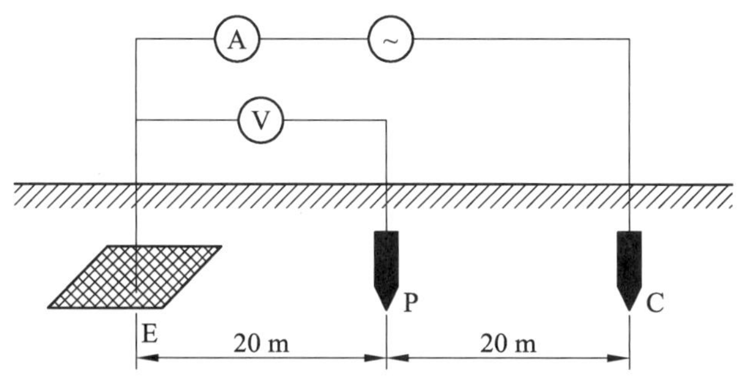

图 2-26　接地电阻测量线路原理

在进行测量时，为防止外界杂散干扰和把辅助接地极的电阻包括在内，一般采用两个电极，一个是把电流引入地下，称为电流极 C，与被测接地体相距较远；另一个用来测量电压，称为电位极 P，测量时 E、P、C 三极必须在一条直线上。

一、接地电阻的测量原理

大地之所以能够导电是因为土壤中含有电解质。如果测量接地电阻时施加的是直流电压，则会引起化学极化作用，使测量结果产生很大的误差，因此测量接地电阻时不能用直流电压，一般都用交流电压。

如图 2-27 所示是用补偿法测量接地电阻的原理电路。图中 E 为接地电极，P 为电位辅助电极，C 为电流辅助电极。E 接接地体，P、C 分别接电位探测针和电流探测针，三者应在一条直线上，间距不小于 20 m。被测接地电阻就是 E、P 之间的土壤散流电阻，不包括电流辅助电极 C 的接地电阻。

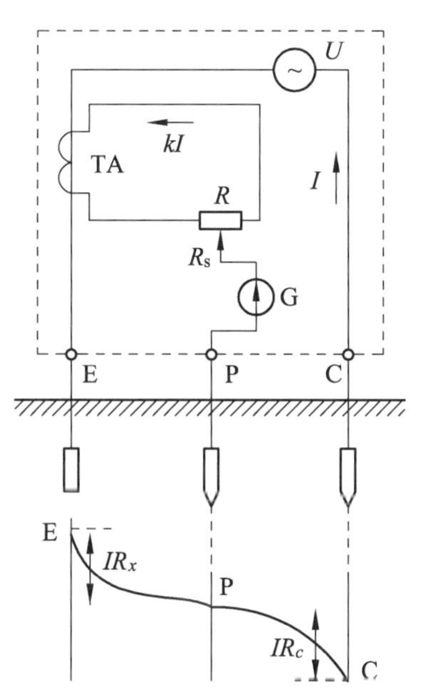

图 2-27　用补偿法测量接地电阻的原理电路图

交流电源的输出电流经电流互感器 TA 的一次绕组到接地电极 E 通过大地和电流辅助探针、电流辅助电极 C 构成闭合回路，在接地电阻 R_x 上形成电压降 IR_x，IR_x 的电位分布如图 2-27

所示，电流互感器 TA 的二次绕组感应电流 kI，并经电位器 R 构成回路，电位器左端电压降为 kIR_s。当检流计指针偏转时，调节电位器使检流计指针为零，则此时有：

$$IR_x = kIR_s$$

$$R_x = \frac{kI}{I} R_s = kR_s \tag{2-9}$$

k 是互感器 TA 的变比。可见，被测接地电阻 R_x 的测量值仅由电流互感器变比和电位器的电阻 R_s 决定，而与辅助电极的接地电阻无关。

三、接地电阻表

接地电阻表主要用于直接测量各种接地装置的接地电阻。其基本结构由手摇发电机、电流互感器、调节电位器及检流计等组成，全部机构装在铝合金铸造的携带式外壳内，附件由两根探测针及三根连接导线。

接地电阻表型号很多，常用的有 ZC-8 型、ZC-29 型等。

（一）ZC-8 型接地电阻表的结构

ZC-8 型接地电阻表是按补偿法的原理制成的，内附手摇交流发电机作为电源，其外形和内部原理电路如图 2-28 所示。

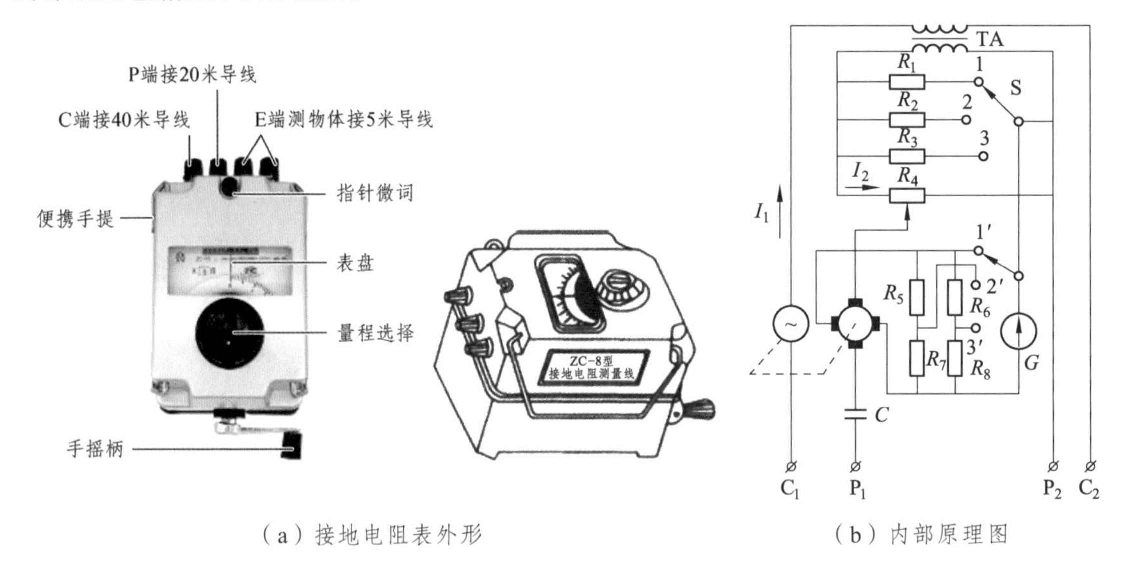

（a）接地电阻表外形　　（b）内部原理图

图 2-28　ZC-8 型接地电阻表

该测量仪有 3 个和 4 个端钮两种。其中，有 4 个端钮的表，一般应将 P_2、C_2 短接后再接到被测接地体，而 3 个端钮的测量仪通常已在内部将 P_2、C_2 短接，再引出一个 E 端钮，测量时直接将 E 接到接地体即可。端钮 P_1、C_1 分别接电位探测针和电流探测针，两探针按要求的距离插入地中。为了扩大测量仪的量程，电路中接有 3 组不同的分流电阻 $R_1 \sim R_3$ 和 $R_5 \sim R_8$，用来实现对电流互感器二次电流以及检流计支路的分流。分流电阻的切换利用联动的转换开关 S 同时进行。对应于转换开关的 3 个挡位，可以得到 0 ~ 1 Ω、0 ~ 10 Ω、0 ~ 100 Ω 3 个量程。

当转换开关置于 1 挡时 $I_1=I_2$，$K=1$；当转换开关置于 2 挡时，$I_2=\dfrac{I_1}{10}$，$k=\dfrac{1}{10}$；当转换开关置于 3 挡时，$I_2=\dfrac{I_1}{100}$，$k=\dfrac{1}{100}$。电位器的旋钮在测量仪的面板上，并带有读数盘。调节电位器使检流计指针指零，则被测接地电阻的值为 $R_X=kR_S$。

（二）ZC-8 型接地电阻表测量接地电阻

测量前，首先将两根探测针分别插入地中，如图 2-29 所示，使被测接地极 E′、电位探测针 P′和电流探测针 C′ 3 点在一条直线上，E′至 P′的距离为 20 m，E′至 C′的距离为 40 m，然后用专用线分别将 E′、P′和 C′接到仪表相应的端钮上。

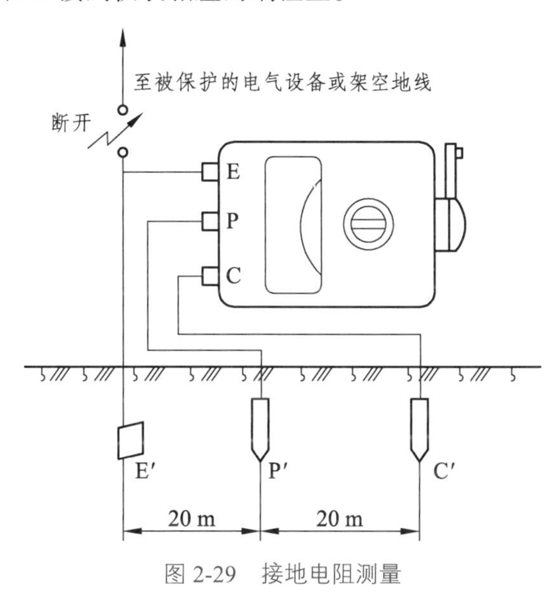

图 2-29　接地电阻测量

测量时，先把仪表放在水平位置，检查检流计的指针是否指在红线上，若不在红线上，则可用调零螺丝进行调零，然后将仪表的倍率标度置于最大倍数，转动发电机手柄，同时调整测量标度盘，使指针位于红线上。如果测量标度盘的读数小于 1，则应将倍率标度置于较小的倍数，再重新调整测量标度盘，以得到正确的读数。

当指针完全平衡在红线上以后，用测量标度盘的读数乘以倍率标度，即为所测的接地电阻值。

使用接地电阻测量仪时，应注意以下两点：

（1）当检流计的灵敏度过高时，可将电位探测针 P′插入土中浅一些；当检流计灵敏度不够时，可在电位探测针 P′和电流探测针 C′周围注水使其湿润。

（2）测量时，应先拆开接地线与被保护设备或线路的连接点，以便得到准确的测量数据。在断开连接点时，应戴绝缘手套。

Task Ⅳ Inspection and use of grounding resistance meter

The grounding in the electric power system is generally divided into three types according to their different functions, namely working grounding, protective grounding, and lightning protection grounding. The grounding of a certain point in the system to ensure the reliable operation of electrical equipment is called working grounding. When the electrical equipment is in operation, the insulation of the electrical equipment may be charged due to breakdown and leakage, which endangers the safety of the person and equipment, so it is generally required to ground the housing of the electrical equipment, which is called protective grounding. In order to prevent lightning attacks, lightning arresters are installed on electrical equipment or transmission lines, and these lightning arresters must also be reliably grounded, which is called lightning protection grounding. In the above grounding systems, the grounding resistance is directly related to the safety of personnel and equipment, and is related to factors such as the structure of the ground, the resistivity of the soil, the geometric dimension and shape of the grounding body, etc. The standard requirements for grounding resistance of various electrical equipment and transmission lines with different voltage classes are specified in the regulations. If the grounding resistance does not meet the requirements, it can not only not guarantee safety, but also cause the illusion of safety and the hidden danger of accidents. Therefore, it is necessary to regularly measure the grounding resistance.

Ⅰ. Concepts of grounding and grounding resistance

The so-called grounding is to use metal conductors to connect the parts of electrical equipment and transmission lines that need to be grounded with metal grounding bodies buried in the soil. The grounding resistance of the grounding body includes the resistance of the grounding body itself, the resistance of the grounding wire, the contact resistance between the grounding body and the ground, and the stray current resistance of the ground. Because the first three resistance is so small that they can be ignored, the grounding resistance generally refers to the stray current resistance.

When there is voltage on the grounding body, there is current flowing into the ground from the grounding body and spreading around, as shown in Fig. 2-25. The closer it is to the grounding body, the smaller the cross section through which the current passes, the greater the resistance, the higher the current density, and the higher the ground potential; The farther away from the grounding body, the larger the cross section through which the current passes, the smaller the resistance, the smaller the current density, and the lower the potential. About 20 m away from the grounding body, the current density is almost zero, and the potential is close to zero, so the grounding resistance is mainly the resistance from the grounding body to the zero-potential point. It is equal to the ratio of the voltage to the ground of the grounding body to the grounding current flowing into the ground through the grounding body $\left(R=\dfrac{U}{I}\right)$. The voltage to the ground is the potential difference between the grounding point of the electrical equipment and the zero potential of the ground.

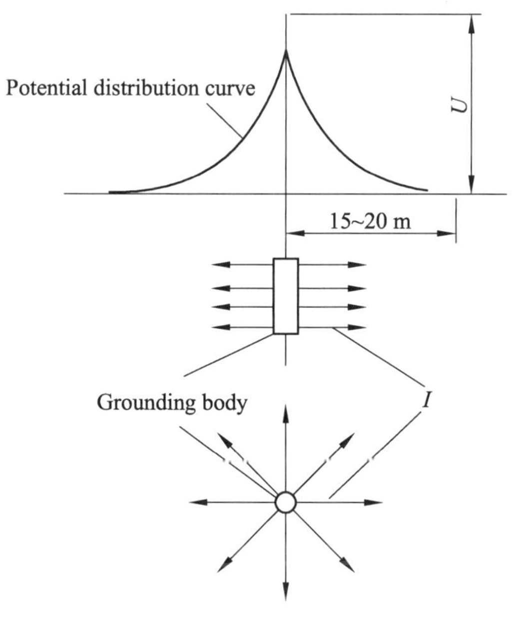

Fig. 2-25 Grounding current and potential distribution

The schematic diagram of the grounding resistance measuring circuit is as shown in Fig. 2-26. Dozens of meters away from the grounding body under test E, the auxiliary grounding electrode C is inserted into the ground, and the AC voltage is applied to the E and C terminals, so that the current I passes through the electrode and the ground, and the current route from the grounding body E is scattered in different directions. The farther away from the E pole, the lower the current density. Because the farther away from the grounding body E, the lower the resistance, and the closer to the grounding body E, the greater the resistance, so most of the voltage falls in the zone near the grounding body.

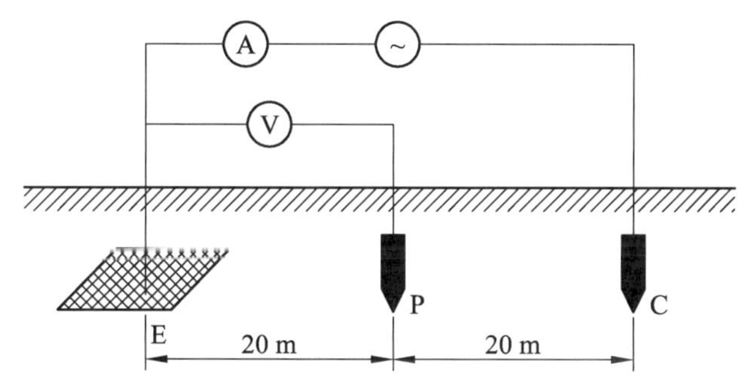

Fig. 2-26 Schematic diagram of grounding resistance measuring circuit

In order to prevent external stray interference and inclusion of the resistance of the auxiliary grounding electrode, two electrodes are generally used in the measurement. One is to introduce the current into the ground, which is called the current electrode C, which is far away from the grounding body to be measured; The other is used to measure voltage, which is called potential electrode P. The three electrodes E, P, and C must be in a straight line when measuring.

Ⅱ. Measurement principle of grounding resistance

The ground conducts electricity because the soil contains electrolytes. If a DC voltage is applied when measuring the grounding resistance, it will cause chemical polarization and cause significant errors in the measurement results. Therefore, DC voltage cannot be used when measuring the grounding resistance, and AC voltage is generally used.

Fig. 2-27 shows the schematic circuit for measuring grounding resistance by compensation method. In the figure, E is the grounding electrode, P is the potential auxiliary electrode, and C is the current auxiliary electrode. E is connected to the grounding body, P and C are connected to the potential probe and the current probe respectively, the three shall be in a straight line, the distance between them is not less than 20 m. The measured grounding resistance R_x is the soil stray resistance between E and P, excluding the grounding resistance of the current auxiliary electrode C.

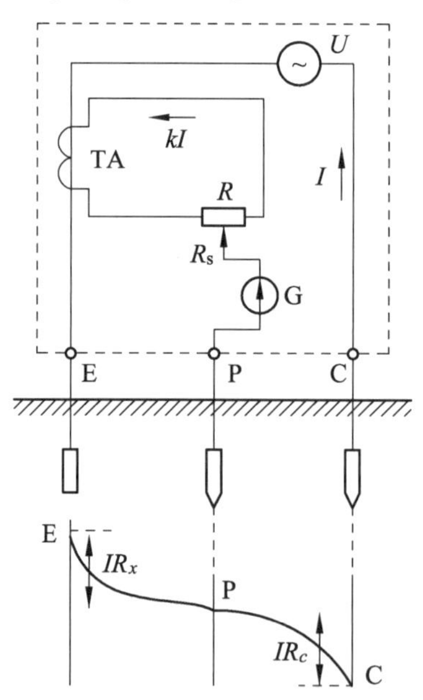

Fig. 2-27 Schematic circuit diagram of measuring grounding resistance by compensation method

The output current I of AC power supply passes through the primary winding of the current transformer TA to the grounding electrode E, forming a closed circuit through the ground and the current auxiliary probe, current auxiliary electrode C. Voltage drop IR_x is formed on the grounding resistance R_x. The potential distribution of IR_x is as shown in Fig. 2-27. The secondary winding of the current transformer TA induces the current W, and a circuit is formed by the potentiometer, and the voltage drop at the left end of the potentiometer is kIR_s. When the galvanometer pointer deflects, adjust the potentiometer so that the galvanometer pointer is zero, then there are:

$$IR_x=kIR_s$$

$$R_x = \frac{kI}{I} R_s = kR_s \tag{2-9}$$

k is the transformation ratio of mutual inductor TA. It can be seen that the measured value of the grounding resistance R_x is only determined by the transformer ratio of the current transformer and the resistance of the potentiometer, but has nothing to do with the grounding resistance of the auxiliary electrode.

Ⅲ. Grounding resistance meter

The grounding resistance meter is mainly used to directly measure the grounding resistance of various grounding devices. Its basic structure consists of a hand generator, a current transformer, an adjusting potentiometer, and a galvanometer, all of which are installed in a portable housing cast by aluminum alloy, with two probes and three connecting conductors attached.

There are many models of grounding resistance meters, such as ZC-8 and ZC-29.

(I) Structure of ZC-8 grounding resistance meter

ZC-8 grounding resistance meter is made according to the principle of compensation method, with hand AC generator as power supply. Its outline and internal schematic circuit are as shown in Fig. 2-28.

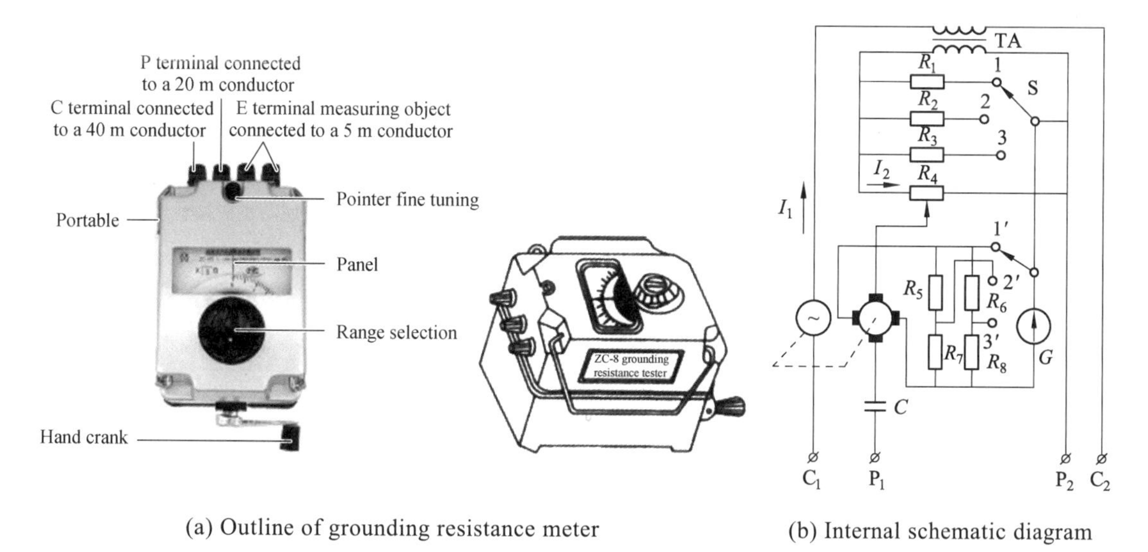

(a) Outline of grounding resistance meter (b) Internal schematic diagram

Fig. 2-28 ZC-8 grounding resistance meter

This tester comes in two types: three and four terminals. For a meter with four terminals, P_2 and C_2 shall generally be short circuited and then connected to the grounding body under test. A meter with three terminals usually has P_2 and C_2 short circuited internally, and then an E terminal is led. When measuring, E can be directly connected to the grounding body. The terminals P_1 and C_1 are connected with the potential probe and the current probe respectively, and the two probes are inserted into the ground at the required distance. In order to expand the measuring range of the tester, three groups of different shunt resistors R_1–R_3 and R_5–R_8 are connected in the circuit to realize the shunt of the secondary current of the current transformer and the branch of the galvanometer. The switching of the shunt resistors is carried out at the same time by using the linked changeover switch S. Corresponding to the three gears of the changeover switch, three ranges of 0–1 Ω, 0–10 Ω, and 0–100 Ω can be obtained. When the changeover switch is in gear 1, $I_2=I_1$, $k=1$; When the changeover switch is in gear 2, $I_2=\frac{I_1}{10}$, $k=\frac{1}{10}$; When the changeover switch is in gear 3, $I_2=\frac{I_1}{100}$, $k=\frac{1}{100}$. The knob of the potentiometer is on the panel of the tester and is equipped with a reading dial. If the potentiometer is adjusted so that the pointer of the galvanometer points to zero, the value of the measured grounding resistance is $R_x=kR_s$.

(Ⅱ) Measuring grounding resistance with a ZC-8 grounding resistance meter

Before the measurement, first insert two probes into the ground respectively, as shown in Fig. 2-29, so that the tested grounding electrode E', potential probe P' and current probe C' are on a straight line, the distance from E' to P' is 20 m, the distance from E' to C' is 40 m, and then the E', P, and C' are connected to the corresponding terminals of the instrument with special wires.

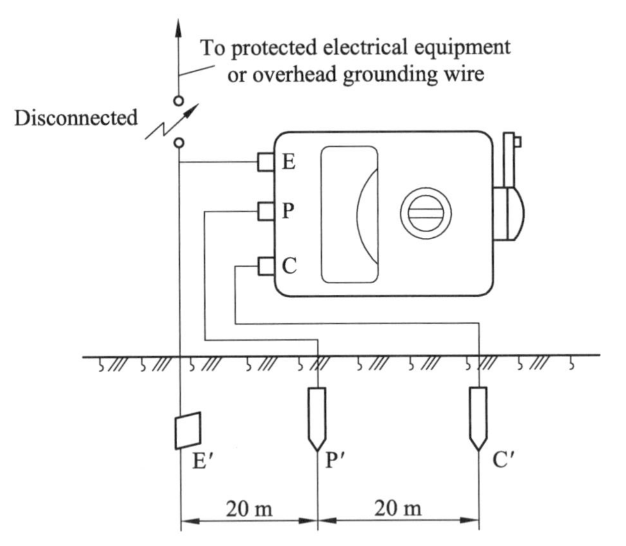

Fig. 2-29 Grounding resistance measurement

During measurement, first put the instrument in a horizontal position, check whether the pointer of the galvanometer is on the red line, if it is not on the red line, use the zeroing screw to zero, then put the magnification scale of the instrument at the maximum multiple, turn the generator handle, and adjust the measurement scale plate so that the pointer is on the red line. If the reading of the measurement scale plate is less than 1, the magnification scale shall be placed in a smaller multiple, and then the measurement scale plate shall be readjusted to get the correct reading.

When the pointer is fully balanced on the red line, the measured grounding resistance is obtained by multiplying the reading of the measurement scale plate by the magnification scale.

When using a grounding resistance tester, the following two points shall be noted:

(1) When the sensitivity of the galvanometer is too high, the potential probe P' can be inserted into the soil shallower. When the sensitivity of the galvanometer is not enough, water can be injected around the potential probe P 'and current probe C' to wet them.

(2) During measurement, the connection point between the grounding wire and the protected equipment or line shall be disconnected first in order to obtain accurate measurement data. When disconnecting the connection point, insulating gloves shall be worn.

任务五　电力测量仪表的检查与使用实训

一、作业任务

5 人一组，正确检查万用表（模拟式和数字式）、钳形电流表、绝缘电阻表和接地电阻表的外观；根据教师给定的测量任务，遵循安全操作要求，按照操作步骤正确使用仪表进行测量。

（1）按给定的测量任务，选择合适的电工仪表。

（2）对所选的仪器仪表进行检查。

（3）正确使用仪器仪表。

（4）正确读数，并对测量数据进行判断。

二、引用标准及文件

（1）《国家电网公司电力安全工作规程（变电部分）》。

（2）《特种作业（电工）安全技术培训大纲和考核标准》。

三、作业条件

在检查和使用仪表时，作业人员精神状态良好，熟悉工作中安全措施、技术措施以及现场工作危险点。

四、作业前准备

1. 现场作业基本要求及条件

勘察现场仪器仪表及设备情况，查阅相关技术资料，包括历史数据及相关规程。

2. 工器具及材料选择

仪器仪表准备：工作台摆放好作业用电工仪器仪表，相关测试线、夹钳、备用电池以及被测量设备，中性笔 1 支、测试结果记录表 1 张。

3. 危险点及预防措施

（1）使用不当造成仪器仪表损坏。

危险点：不正确使用仪器仪表，造成仪器仪表损坏。

预防措施：正确使用仪器仪表。

（2）作业人员伤害。

预防措施：正确按照安全措施要求进行操作。

4. 作业人员分工

现场工作负责人（监护人）：×××

现场作业人员：×××

五、作业规范及要求

（1）现场操作时，由老师给出检查及使用仪器仪表要求，学员按现场规程模拟演示。

（2）万用表（模拟式和数字式）、钳形电流表、绝缘电阻表和接地电阻表中抽选 2 种仪器仪表进行。

（3）必须按操作规程严格执行，一旦发生异常立刻停止并报告。

六、作业流程及标准（表 2-5）

表 2-5 电工仪器仪表安全使用流程及考核评分标准

班级		姓名学号	考评员成绩			
序号	作业名称	质量标准	分值/分	扣分标准	扣分	得分
1	选用合适的电工仪表	正确口述各种电工仪表的作用，针对教师布置的测量任务，正确选择合适的电工仪表（万用表、钳形电流表、绝缘电阻表、接地电阻表）	20	口述各种电工仪表的作用，不正确扣 10 分； 针对考评员布置的测量任务，正确选择合适的电工仪表（万用表、钳形电流表、绝缘电阻表、接地电阻测试仪），仪表选择不正确扣 10 分		
2	仪表检查	正确检查仪表的外观有无破损，有无电量充足的电池，摇臂是否正常转动，指针是否正常偏转，仪表接线端钮是否完好	20	正确检查仪表的外观，未检查外观扣 10 分； 未检查完好性，扣 10 分		
3	正确使用仪表	遵循安全操作要求，按照操作步骤正确使用仪表	50	遵循安全操作要求，按照操作步骤正确使用仪表，得 50 分； 操作步骤违反安全规程得零分，操作步骤不完整视情况扣 5～50 分		
4	对测量结果进行判断	能正确对测量的结果进行分析判断	10	未能对测量的结果进行分析判断，扣 10 分		
合计			100			

Task Ⅴ Practical training of inspection and use of electric power measuring instrument

Ⅰ. Operating tasks

In a group of five persons, check the appearance of multimeter (analog and digital), clip-on ammeter, insulation resistance meter, and grounding resistance meter correctly; According to the measurement tasks given by the instructor, follow the requirements of safe operation, and correctly use the instrument to measure according to the operation steps.

(1) Select the appropriate electrical instrument according to the given measurement tasks.

(2) Check the selected instruments.

(3) Correctly use instruments.

(4) Read correctly and make judgments on the measured data.

Ⅱ. Referenced standards and documents

(1) *Electric Power Safety Working Regulations (Power Transformation) of State Grid Corporation of China*;

(2) *Safety Technology Training Outline and Appraisal Standards for Special Operation (Electrician)*.

Ⅲ. Operating conditions

When checking and using the instrument, the operation personnel shall be in a good mental state and familiar with safety measures, technical measures, and dangerous points of field work.

Ⅳ. Preparation before operation

1. Basic requirements and conditions for on-site operations

Conduct a survey of instruments and equipment on site and refer to relevant technical data, including historical data and related procedures.

2. Selection of tools and instruments and materials

Preparation of instruments: Place electrical instruments, related test lines, clamps, spare batteries, and equipment to be measured on the workbench. One gel pen and one test result record sheet.

3. Dangerous points and preventive measures

(1) Improper use causes damage to instruments.

Dangerous points: Incorrect use of instruments, resulting in instrument damage.

Prevention and control measures: Use instruments correctly.

(2) Injury to the operator.

Prevention and control measures: Operate correctly according to the requirements of safety measures.

4. Division of labor among operators

Person in charge of on-site work (supervisor): ×××

On-site operator: ×××

Ⅴ. Operating specifications and requirements

(1) When it comes to on-site operation, the instructor shall specify requirements for inspection and use of instruments, and the trainees shall simulate the demonstration according to the on-site procedures.

(2) Select two instruments from the multimeter (analog and digital), clip-on ammeter, insulation resistance meter, and grounding resistance meter.

(3) The operating procedures must be strictly followed. Stop immediately and report in the event of any anomaly.

Ⅵ. Operation processes and standards (see Tab. 2-5)

Tab. 2-5 Safe use processes and assessment and scoring standards of electrical instruments

Class		Name		Student ID		Examiner		Score	
S/N	Operation name	Quality standard			Points	Deduction criteria		Deduction	Score
1	Select appropriate electrical instruments	Correctly verbally describe the functions of various electrical instruments. Correctly select appropriate electrical instruments (multimeter, clip-on ammeter, insulation resistance meter, grounding resistance meter) according to the measurement tasks assigned by the instructor			20	Verbally describe the functions of various electrical instruments. Deduct 10 points for incorrect verbal description; For measurement tasks assigned by the evaluator, correctly select appropriate electrical instruments (multimeter, clip-on ammeter, insulation resistance meter, grounding resistance meter). Deduct 10 points for incorrect instrument selection			
2	Instrument inspection	Correctly check whether the outline of the instrument is damaged, whether the battery is fully charged, whether the rocker arm rotates normally, whether the pointer is deflected normally, and whether the instrument terminal is in good condition			20	Check the appearance of the instrument correctly. Deduct 10 points if failing to check the appearance; Deduct 10 points if failing to check the integrity			
3	Correctly use instruments	Follow the safe operation requirements and use the instrument correctly according to the operating steps			50	Get 50 points for following the safe operation requirements and using the instrument correctly according to the operating steps, Get 0 point for violating safety regulations in operating steps. Deduct 5-50 points for incomplete operating steps as appropriate			

Continued

S/N	Operation name	Quality standard	Points	Deduction criteria	Deduction	Score
4	Judge the measurement results	Be able to correctly analyze and judge the measurement results	10	Deduct 10 points for failing to analyze and judge the measurement results		
Total			100			

模块 三 安全组织措施的应用

在电力网络建设、运行和维护工作中，为防止事故发生，保障现场作业人员的安全，必须严格执行《国家电网公司电力安全工作规程》的规定。在作业的整个过程中，必须按照规定完成保证作业人员安全的组织措施和技术措施。

其中，应遵守的组织措施有现场勘察制度、工作票制度、工作许可制度、工作监护制度与工作间断、转移和终结制度。

学习目标

（1）熟悉《国家电网公司电力安全工作规程（变电部分）》第六节保证安全的组织措施，整体流程如图 3-1 所示。

（2）理解保证安全的组织措施中具体制度的意义。

（3）掌握电气设备现场勘察的要点。

（4）掌握工作票执行的流程。

（5）掌握现场工作许可的流程。

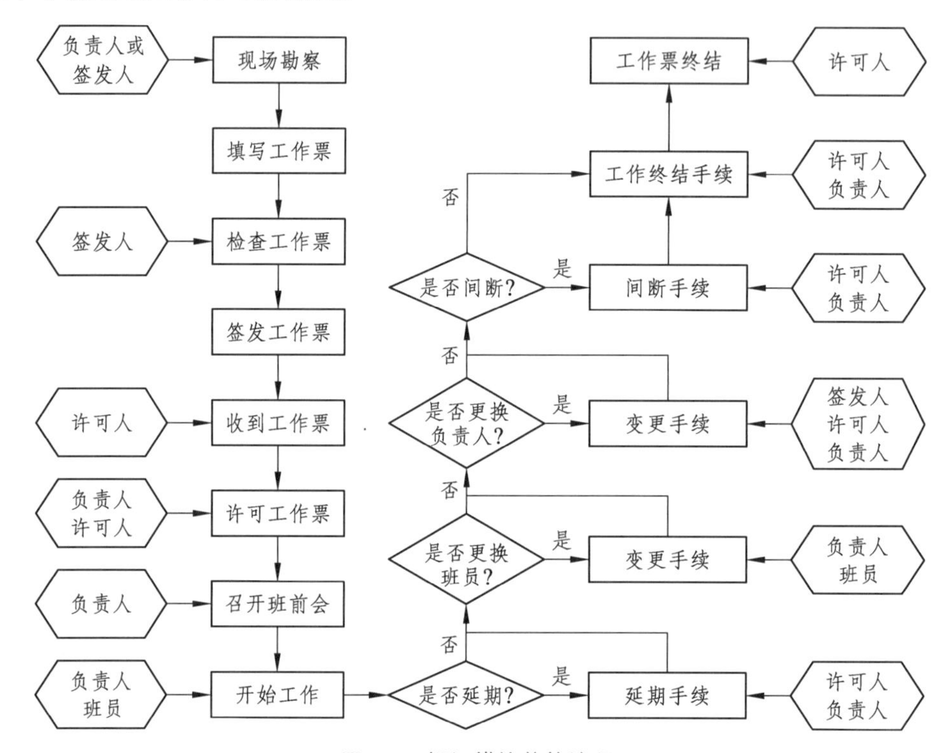

图 3-1　组织措施整体流程

Module Ⅲ Application of safety and organizational measures

In the construction, operation and maintenance of power network, to prevent accidents and ensure the safety of on-site operators, the provisions of the *Electric Power Safety Working Regulations of State Grid Corporation of China* must be strictly implemented. Throughout the entire process of the operation, it is necessary to take the organizational and technical measures required to ensure the operation of personnel in accordance with regulations.

Among them, the organizational measures that shall be followed include on-site survey system, work ticket system, work permit system, work supervision system and work interruption, transfer and termination system.

◉ Learning objectives:

(1) Be familiar with the organizational measures for ensuring safety as described in Section 6 of the *Electric Power Safety Working Regulations (Power Transformation) of State Grid Corporation of China*. The overall process is as shown in Fig. 3-1.

(2) Understand the significance of specific systems in organizational measures for ensuring safety.

(3) Master the key points of on-site survey of electrical equipment.

(4) Master the process of work ticket implementation.

(5) Master the process of on-site work permit.

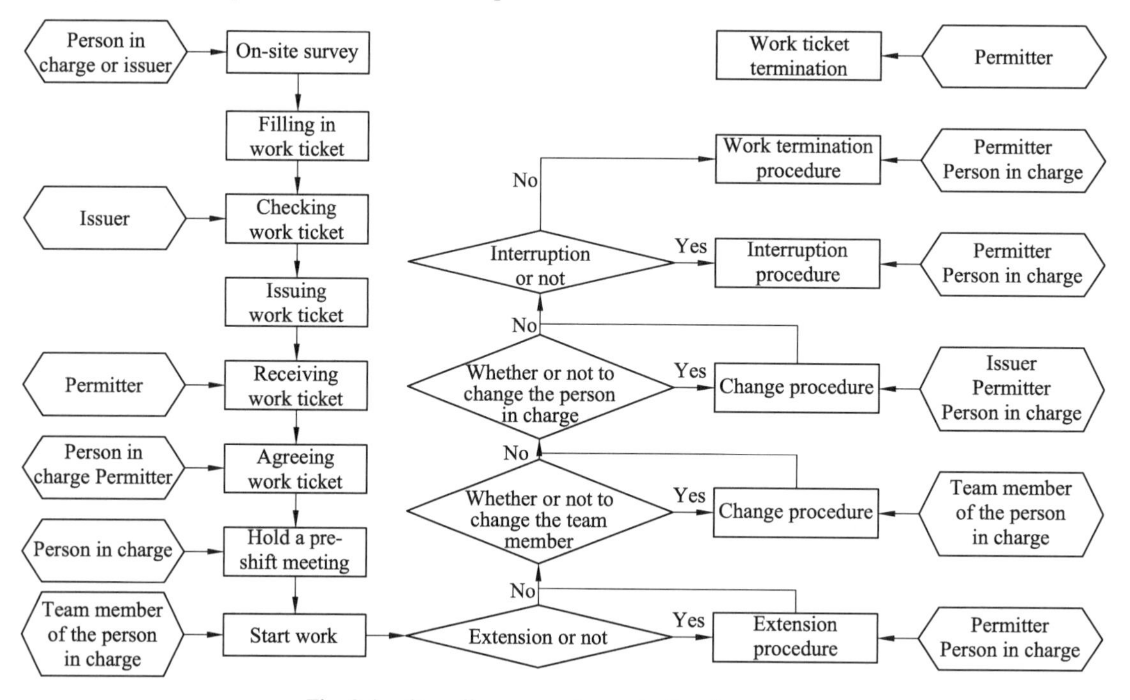

Fig. 3-1 Overall process of organizational measures

任务一　现场勘察制度及其应用

一、现场勘察制度

变电检修（施工）作业，工作票签发人或工作负责人认为有必要现场勘察的，检修（施工）单位应根据作业任务组织现场勘察，并填写现场勘察记录。现场勘察由工作票签发人或工作负责人组织。

二、现场勘察的内容

现场勘察是在正式作业前，由相关负责人率先去现场，了解现场设备、环境等情况，以及清楚现场的危险点。这一环节对于检修工作而言，是必不可少的一环，也是保证人身、电网、设备安全的重要措施，所以现场勘察负责人应当由完全清楚本次工作的具体工作内容的工作票签发人或工作负责人担任。

在现场勘察时，应当在现场勘察记录报告中记录以下几点：

（1）现场勘察的基本情况：包括勘察的单位、班组、人员、时间、地点等信息。

（2）现场设备的双重名称与作业内容：应当在现场勘察记录报告中抄录现场设备的双重名称，详细写出现场作业内容。

（3）现场设备或现场线路的具体情况：应在现场勘察记录报告中详细描述现场设备、线路目前的具体情况，包括但不仅限于线路的跨越、交叉、相关设备的尺寸、带电情况、安全距离等。必要时应当使用完善的图文相互配合的形式来描述。

（4）作业现场需要停电的部位与保留的带电部位：应在现场勘察记录报告中详细描述现场作业时现场带电的设备，作业现场停电的设备以及停电设备上将保留的带电部位。

（5）作业现场的危险点：应在现场勘察记录报告中详细描述在现场作业时，可能发生的所有类型的危险点，包括但不仅限于高压触电、低压触电、高处坠落、物件伤人、油污等化学物质伤人、设备损伤等。

（6）作业现场应当采取的安全措施：根据上述危险点，应在现场勘察记录报告中详细描述作业前应当布置好的安全措施，包括但不仅限于应拉合的隔离开关、断路器，应装设的接地线，应悬挂的标示牌，应装设的围栏，等。

（7）现场作业时应当注意的事项：根据上述危险点以及安全措施，应在现场勘察记录报告中详细描述作业时应当注意的事项。

勘察记录报告中的以上内容在必要时应当使用完善的图文相互配合的形式来描述，以防出错。

Task Ⅰ On-site survey system and its application

Ⅰ. On-site survey system

For the substation maintenance (construction) operations, if the issuer of the work ticket or the person in charge of work considers that it is necessary to carry out the on-site survey, the maintenance (construction) organization shall organize the on-site survey according to the operation tasks and fill in the on-site survey records. The on-site survey shall be organized by the issuer of the work ticket or the person in charge of work.

Ⅱ. Content of on-site survey

On-site survey means that before the formal operation, the relevant person in charge takes the lead to go to the site to understand the on-site equipment, environment, and other conditions, as well as ascertaining the dangerous points on the site. This link is not only an indispensable part of maintenance work, but also an important measure to ensure the safety of people, power grid, and equipment. Therefore, the person in charge of the on-site survey should be the issuer of the work ticket or the person in charge of work who is fully aware of the specific content of this work.

During on-site survey, the following points shall be recorded in the on-site survey record report:

(1) Basic information of on-site survey: including the survey unit, team, members, time, place, etc.

(2) Double names and work content of equipment on site: The double names of equipment on site shall be copied in the on-site survey record report; Write out the content of the on-site operation in detail.

(3) Specific conditions of equipment or lines on site: The current specific conditions of equipment and lines shall be described in detail in the on-site survey record report, including but not limited to spanning and crossing of lines, dimensions, state of electrification, and safe distance of equipment. If necessary, combine text with photos to describe related information.

(4) Parts to which power outage will occur and live parts remained on the work site: Live equipment, equipment to which power outage will occur and live parts remained on the equipment with power outage on the work site shall be described in detail in the on-site survey record report.

(5) Dangerous points on the work site: All kinds of potential dangerous points shall be described in detail in the on-site survey record report, including but not limited to high voltage electric shock, low voltage electric shock, fall from height, personal injury caused by objects, personal injury caused by chemical substances such as oil pollutant, and equipment damage, etc.

(6) Safety measures required on the work site: According to the above dangerous points, safety measures which shall be provided shall be described in detail in the on-site survey record report,

including but not limited to isolating switches and circuit breakers to be closed, grounding wires to be installed, sign boards to be hung and fences to be installed.

(7) Precautions required on the work site: According to the above dangerous points and safety measures, precautions required shall be described in detail in the on-site survey record report.

The above content in the survey record report shall be described in a complete form of graphic and textual coordination to prevent errors.

任务二　工作票制度及其应用

一、工作票的种类

1. 工作票及事故紧急抢修单

在电气设备上的工作，应填用工作票或事故紧急抢修单，其方式有以下 6 种：

（1）填用变电站（发电厂）第一种工作票。

（2）填用电力电缆第一种工作票。

（3）填用变电站（发电厂）第二种工作票。

（4）填用电力电缆第二种工作票。

（5）填用变电站（发电厂）带电作业工作票。

（6）填用变电站（发电厂）事故紧急抢修单。

2. 填用第一种工作票的工作

（1）高压设备上工作需要全部停电或部分停电者。

（2）二次系统和照明等回路上的工作，需要将高压设备停电者或做安全措施者。

（3）高压电力电缆需停电的工作。

（4）换流变压器、直流场设备及阀厅设备需要将高压直流系统或直流滤波器停用者。

（5）直流保护装置、通道和控制系统的工作，需要将高压直流系统停用者。

（6）换流阀冷却系统、阀厅空调系统、火灾报警系统及图像监视系统等工作，需要将高压直流系统停用者。

（7）其他工作需要将高压设备停电或要做安全措施者。

3. 填用第二种工作票的工作

（1）控制盘和低压配电盘、配电箱、电源干线上的工作。

（2）二次系统和照明等回路上的工作，无须将高压设备停电者或做安全措施者。

（3）转动中的发电机、同期调相机的励磁回路或高压电动机转子电阻回路上的工作。

（4）非运维人员用绝缘棒、核相器和电压互感器定相或用钳形电流表测量高压回路的电流。

（5）大于表 3-1 规定距离的相关场所和带电设备外壳上的工作，以及无可能触及带电设备导电部分的工作。

（6）高压电力电缆不需停电的工作。

（7）换流变压器、直流场设备及阀厅设备上工作，无须将直流单、双极或直流滤波器停用者。

（8）直流保护控制系统的工作，无须将高压直流系统停用者。

（9）换流阀水冷系统、阀厅空调系统、火灾报警系统及图像监视系统等工作，无须将高压直流系统停用者。

4. 填用带电作业工作票的工作

带电作业或与邻近带电设备距离小于表 3-1、大于表 3-2 规定的工作。

表 3-1　设备不停电时的安全距离

电压等级/kV	安全距离/m	电压等级/kV	安全距离/m
10 及以下（13.8）	0.70	1 000	8.70
20、35	1.00	±50 及以下	1.50
66、110	1.50	±400	5.90
220	3.00	±500	6.00
330	4.00	±660	8.40
500	5.00	±800	9.30
750	7.20		

注：① 表中未列电压等级按高一挡电压等级安全距离确定。

② 表中±400 kV 数据是按海拔 3 000 m 校正的，海拔 4 000 m 时安全距离为 6.00 m。750 kV 数据是按海拔 2 000 m 校正的，其他等级数据按海拔 1 000 m 校正。

表 3-2　带电作业时人身与带电体间的安全距离

电压等级/kV	10	35	66	110	220	330	500	750	1000	±400	±500	±660	±800
距离/m	0.4	0.6	0.7	1.0	1.8 （1.6）	2.6	3.4 （3.2）	5.2 （5.6）	6.8 （6.0）	3.85	3.4	4.5	6.8

注：① 表中数据根据线路带电作业安全要求提出。

② 220 kV 带电作业安全距离因受设备限制达不到 1.8 m 时，经单位分管生产的领导（总工程师）批准，并采取必要的措施后，可采用括号内 1.6 m 的数值。

③ 500 kV 电压等级时，海拔 500 m 以下，500 kV 取 3.2 m 值，但不适用于 500 kV 紧凑型线路。海拔在 500 ~ 1 000 m 时，500 kV 取 3.4 m 值。

④ 750 kV 电压等级时，直线为边相或中相值。5.2 m 为海拔 1 000 m 以下值，5.6 m 为海拔 2 000 m 以下的距离。

⑤ 1 000 kV 电压等级时，此为单回输电线路数据，括号中数据 6.0 m 为边相值，6.8 m 为中相值。表中数值不包括人体占位间隙，作业中需考虑人体占位间隙不得小于 0.5 m。

⑥ 表中土 400 kV 数据是按海拔 3 000 m 校正的，海拔为 3 500 m、4 000 m、4 500 m、5 000 m、5 300 m 时最小安全距离依次为 3.90 m、4.10 m、4.30 m、4.40 m、4.50 m。

⑦ 表中± 660 kV 数据是按海拔 500 ~ 1 000 m 校正的；海拔 1 000 ~ 1 500 m、1 500 ~ 2 000 m 时最小安全距离依次为 4.7 m、5.0 m。

5. 填用事故紧急抢修单的工作

事故紧急抢修可不用工作票，但应使用事故紧急抢修单。

非连续进行的事故修复工作，应使用工作票。

6. 其他情况

运维人员实施不需高压设备停电或做安全措施的变电运维一体化业务项目时，可不使用工作票，但应以书面形式记录相应的操作和工作等内容。

各单位应明确发布所实施的变电运维一体化业务项目及所采取的书面记录形式。

二、工作票的填写与签发

（一）工作票的填写、打印规范

（1）工作票应使用黑色或蓝色的钢（水）笔或圆珠笔填写与签发，一式两份，内容应正确，填写应清楚，不得任意涂改。如有个别错、漏字需要修改，应使用规范的符号，字迹应清楚。

（2）用计算机生成或打印的工作票应使用统一的票面格式，由工作票签发人审核无误，手工或电子签名后方可执行。工作票一份应保存在工作地点，由工作负责人收执；另一份由工作许可人收执，按值移交。工作许可人应将工作票的编号、工作任务、许可及终结时间记入登记簿。

一般情况是仅允许签发人手工签名后执行。所谓“按值移交”是指依据交接班制度移交工作票。

（二）填写、签发资格

（1）一张工作票中，工作许可人与工作负责人不得互相兼任。若工作票签发人兼任工作许可人或工作负责人，应具备相应的资质，并履行相应的安全责任。

（2）工作票由工作负责人填写，也可以由工作票签发人填写。

一般是由工作负责人负责填写工作票，无论是签发人还是负责人填写工作票，均需在签发人检查、签字后，才能生效。

（3）工作票由设备运维管理单位（部门）签发，也可由经设备运维管理单位（部门）审核合格且经批准的检修及基建单位签发。检修及基建单位的工作票签发人及工作负责人名单应事先送有关设备运维管理单位（调度控制中心）备案。

（4）承发包工程中，工作票可实行“双签发”形式。签发工作票时，双方工作票签发人在工作票上分别签名，各自承担本部分工作票签发人相应的安全责任。

（5）供电单位或施工单位到用户变电站内施工时，工作票应由有权签发工作票的供电单位、施工单位或用户单位签发。

（三）总分工作票

（1）第一种工作票所列工作地点超过两个，或有两个及以上不同的工作单位（班组）在一起工作时，可采用总工作票和分工作票。总、分工作票应由同一个工作票签发人签发。总工作票上所列的安全措施应包括所有分工作票上所列的安全措施。几个班同时进行工作时，总工作票的工作班成员栏内，只填各分工作票的负责人，不必填写全部工作班人员姓名。分工作票上要填写工作班人员姓名。

（2）总、分工作票在格式上与第一种工作票一致。

（3）分工作票应一式两份，由总工作票负责人和分工作票负责人分别收执。分工作票的许可和终结，由分工作票负责人与总工作票负责人办理。分工作票应在总工作票许可后才可许可；总工作票应在所有分工作票终结后才可终结。

总分工作票的签发人为同一个人，总工作票的负责人为分工作票的许可人，总工作票的工作班成员为分工作票的负责人。在分工作票实施时，其中一张由总工作票负责人收执，一张由分工作票负责人收执。

三、工作票的使用

（1）一个工作负责人不能同时执行多张工作票，工作票上所列的工作地点，以一个电气连接部分为限。

① 所谓一个电气连接部分是指电气装置中，可以用隔离开关（刀闸）同其他电气装置分开的部分。

② 直流双极停用、换流变压器及所有高压直流设备均可视为一个电气连接部分。

③ 直流单极运行、停用极的换流变压器、阀厅、直流场设备、水冷系统可视为一个电气连接部分。双极公共区域为运行设备。

（2）一张工作票上所列的检修设备应同时停、送电，开工前工作票内的全部安全措施应一次完成。若至预定时间，一部分工作尚未完成，需继续工作而不妨碍送电者，在送电前，应按照送电后现场设备带电情况，办理新的工作票，布置好安全措施后，方可继续工作。

（3）若以下设备同时停、送电，可使用同一张工作票：

① 属于同一电压等级、位于同一平面场所，工作中不会触及带电导体的几个电气连接部分。

② 一台变压器停电检修，其断路器（开关）也配合检修。

③ 全站停电。

（4）同一变电站内在几个电气连接部分上依次进行不停电的同一类型的工作，可以使用一张第二种工作票。

（5）在同一变电站内，依次进行的同一类型的带电作业可以使用一张带电作业工作票。

（6）持线路或电缆工作票进入变电站或发电厂升压站进行架空线路、电缆等工作，应增填工作票份数，由变电站或发电厂工作许可人许可，并留存。上述单位的工作票签发人和工作负责人名单应事先送有关运维单位备案。

（7）需要变更工作班成员时，应经工作负责人同意，在对新的作业人员进行安全交底手续后，方可进行工作。非特殊情况不得变更工作负责人，如确需变更工作负责人应由工作票签发人同意并通知工作许可人，工作许可人将变动情况记录在工作票上。工作负责人允许变更一次。原、现工作负责人应对工作任务和安全措施进行交接。

工作班成员变更后，工作票中原有“工作班成员”栏不进行改变，仅在“工作人员变动情况”一栏添加；新到的工作班成员，需要进行交底手续，并且在“确认工作负责人布置的工作任务和安全措施”栏目签字后，方可工作。

工作负责人为现场工作中最重要的角色，变更人选将会带来一定安全隐患，所以非特殊情况不得变更。若工作负责人变更一次之后，新负责人因不可抗力不能继续负责现场工作，则须提前结束该工作票，由其他负责人重新办理工作票手续，方可重新开始工作。

（8）在原工作票的停电及安全措施范围内增加工作任务时，应由工作负责人征得工作票签发人和工作许可人同意，并在工作票上增填工作项目。若需变更或增设安全措施者应填用新的工作票，并重新履行签发许可手续。

（9）变更工作负责人或增加工作任务，如工作票签发人（和工作许可人）无法当面办理，应通过电话联系，并在工作票登记簿和工作票上注明。

（10）第一种工作票应在工作前一日送达运维人员，可直接送达或通过传真、局域网传送，但传真传送的工作票许可应待正式工作票到达后履行。临时工作可在工作开始前直接交给工作许可人。第二种工作票和带电作业工作票可在进行工作的当天预先交给工作许可人。

（11）工作票有破损不能继续使用时，应补填新的工作票，并重新履行签发许可手续。

四、工作票的有效期与延期

（1）第一、二种工作票和带电作业工作票的有效时间，以批准的检修期为限。

（2）第一、二种工作票需办理延期手续，应在工期尚未结束以前由工作负责人向运维负责人提出申请（属于调控中心管辖、许可的检修设备，还应通过值班调控人员批准），由运维负责人通知工作许可人给予办理。第一、二种工作票只能延期一次。带电作业工作票不准延期。

五、变电“五种人”

（一）工作票所列人员的基本条件

（1）1 作票签发人应是熟悉人员技术水平、熟悉设备情况、熟悉本规程，并具有相关工作经验的生产领导人、技术人员或经本单位批准的人员。工作票签发人员名单应公布。

（2）工作负责人（监护人）应是具有相关工作经验，熟悉设备情况和本规程，经工区（车间，下同）批准的人员。工作负责人还应熟悉工作班成员的工作能力。

（3）工作许可人应是经工区批准的有一定工作经验的运维人员或检修操作人员（进行该工作任务操作及做安全措施的人员）；用户变、配电站的工作许可人应是持有效证书的高压电气工作人员。

（4）专责监护人应是具有相关工作经验，熟悉设备情况和本规程的人员。

（二）工作票所列人员的安全责任

1. 工作票签发人

（1）确认工作必要性和安全性。

（2）确认工作票上所填安全措施是否正确完备。

（3）确认所派工作负责人和工作班人员是否适当和充足。

2. 工作负责人（监护人）

（1）正确安全地组织工作。

（2）检查工作票所列安全措施是否正确完备，是否符合现场实际条件，必要时予以补充。

（3）工作前对工作班成员进行工作任务、安全措施、技术措施交底和危险点告知，并确认每个工作班成员都已签名。

（4）严格执行工作票所列安全措施。

（5）监督工作班成员遵守本规程，正确使用劳动防护用品和安全工器具以及执行现场安全措施。

（6）关注工作班成员身体状况精神状态是否出现异常迹象，人员变动是否合适。

3. 工作许可人

（1）负责审查工作票所列安全措施是否正确、完备，是否符合现场条件。

（2）工作现场布置的安全措施是否完善，必要时予以补充。

（3）负责检查检修设备有无突然来电的危险。

（4）对工作票所列内容即使发生很小疑问，也应向工作票签发人询问清楚，必要时应要求作详细补充。

4. 专责监护人

（1）明确被监护人员和监护范围。

（2）工作前对被监护人员交待监护范围内的安全措施，告知危险点和安全注意事项。

（3）监督被监护人员遵守本规程和现场安全措施，及时纠正被监护人员的不安全行为。

5. 工作班成员

（1）熟悉工作内容、工作流程，掌握安全措施，明确工作中的危险点，并在工作票上履行交底签名确认手续。

（2）服从工作负责人（监护人）、专责监护人的指挥，严格遵守本规程和劳动纪律，在确定的作业范围内工作，对自己在工作中的行为负责，互相关心工作安全。

（3）正确使用施工器具、安全工器具和劳动防护用品。

变电“五种人”中，“工作票签发人”“工作负责人”“工作许可人”这三者也称为“三种人”。

Task Ⅱ Work ticket system and its application

Ⅰ. Types of work tickets

1. Work ticket and emergency repair order

For work on electrical equipment, a work ticket or emergency repair order shall be filled out, which covers the following 6 types:

(1) Fill out the first type of work ticket for substation (power plant).

(2) Fill out the first type of work ticket for power cable.

(3) Fill out the second type of work ticket for substation (power plant).

(4) Fill out the second type of work ticket for power cable.

(5) Fill out the live-wire work ticket for substation (power plant).

(6) Fill out the emergency repair order for substation (power plant).

2. Work requiring the first type of work ticket

(1) Work on high voltage equipment requires a total or partial power outage.

(2) Work on secondary systems and lighting circuits requires the power outage of high-voltage equipment or the adoption of safety measures.

(3) Work on high-voltage power cables requires the power outage.

(4) Work on converter transformer, DC field equipment, and valve hall equipment requires disabling the high voltage DC system or DC filter.

(5) Work on DC protection devices, channels, and control systems requires disabling the high voltage DC system.

(6) Work on converter valve cooling system, valve hall air conditioning system, fire alarm system, and image monitoring system requires disabling the high voltage DC system.

(7) Other work that requires the power outage of high-voltage equipment or the adoption of safety measures.

3. Work requiring the second type of work ticket

(1) Work on the control panel and low-voltage distribution panel, distribution box, and power supply main.

(2) Work on secondary systems and lighting circuits, which does not require the power outage of high-voltage equipment or the adoption of safety measures.

(3) Work on the rotating generator, excitation circuit of synchronous phase adjuster, or rotor resistance circuit of high voltage motor.

(4) Non-O&M personnel use insulating rods, phasing testers, and voltage transformers for phasing or use clip-on ammeters to measure the current of the high voltage circuit.

(5) Work on related sites and live equipment housings greater than the distance specified in Tab. 3-1, and the work where it is impossible to touch the conductive part of the live equipment.

(6) Work on high-voltage power cables that does not require the power outage.

(7) Work on converter transformer, DC field equipment, and valve hall equipment that does not require disabling the DC single, bipolar, or DC filter.

(8) Work on DC protection control systems that does not require disabling the high voltage DC system.

(9) Work on converter valve water cooling system, valve hall air conditioning system, fire alarm system, and image monitoring system that does not require disabling the high voltage DC system.

4. Work requiring the live-wire work ticket

Live-wire work or work with a distance from adjacent live equipment less than that specified in Table 3-1 and greater than that specified in Table 3-2.

Tab. 3-1 Safe distance for equipment at charged state

Voltage class/kV	Safe distance/m	Voltage class/kV	Safe distance/m
10 and below (13.8)	0.70	1,000	8.70
20, 35	1.00	± 50 and below	1.50
66, 110	1.50	±400	5.90
220	3.00	±500	6.00
330	4.00	±660	8.40
500	5.00	±800	9.30
750	7.20		

Note: ① When no voltage class is listed in the table, safe distance for higher voltage class shall be applied.
② The ±400 kV data in the table has been adjusted for an altitude of 3,000 m. At an altitude of 4,000 m, the recommended safe distance is 6.00 m. 750 kV data is calibrated based on 2,000 m altitude, while data of other classes is calibrated based on 1,000 m altitude.

Tab. 3-2 Safe distance between human body and a charged body during live-wire work

Voltage class/kV	10	35	66	110	220	330	500	750	1,000	±400	±500	±660	±800
Distance/m	0.4	0.6	0.7	1.0	1.8 (1.6)	2.6	3.4 (3.2)	5.2 (5.6)	6.8 (6.0)	3.85	3.4	4.5	6.8

Note: ① The data in the table is put forward according to the safety requirements of live-wire work
② When the safety distance for 220 kV live-wire work is less than 1.8 m due to equipment limitations, with the approval of the leader in charge of production (chief engineer) and necessary measures taken, the value of 1.6 m in parentheses can be used.
③ When the voltage class is 500 kV and the altitude is below 500 m, the value of 3.2 m is taken for 500 kV, but it is not applicable to 500 kV compact lines. When the altitude is between 500 m to 1,000 m, the value of 3.4m is taken for 500 kV.
④ For the 750 kV voltage class, the side or middle phase value of the tangent tower. 5.2 m is applicable to altitude below 1,000 m, and 5.6 m is applicable to altitude below 2,000 m.
⑤ At a voltage class of 1,000 kV, this is the data for a single circuit transmission line. The data in parentheses 6.0 m is the side phase value and 6.8 m is the middle phase value. The values in the table do not include the human body occupation clearance. In the operation, it shall be considered that the human body occupation clearance shall not be less than 0.5 m.
⑥The ±400 kV data in the table is corrected based on 3,000 m altitude. The minimum safety distances for altitudes of 3,500 m, 4,000 m, 4500 m, 5,000 m, and 5,300 m are 3.90 m, 4.10 m, 4.30 m, 4.40 m, and 4.50 m, respectively.
⑦ The ±660 kV data in the table is corrected based on 500–1,000 m altitude. The minimum safety distances for altitudes of 1,000–1,500 m and 1,500–2,000 m are 4.7 m and 5.0 m, respectively.

5. Work requiring emergency repair order

Emergency repair may not require work ticket, but requires emergency repair order.

Non-continuous accident repair requires work tickets.

6. Other conditions

When O&M personnel implement an integrated O&M project of substation that does not require high voltage equipment power outage or adoption of safety measures, they may not use work tickets, but shall record the corresponding operations and work content in written form.

Each organization shall clearly issue the integrated O&M project of substation and the written record form adopted.

Ⅱ. Filling out and issuing work tickets

(Ⅰ) Specification for filling out and printing work tickets

(1) The work ticket shall be filled out and issued in black or blue pen (fountain pen) or ballpoint pen, shall be in duplicate, the content shall be correct, the filling shall be clear, and no arbitrary alteration is allowed. If there are individual errors or omissions that need to be corrected, standardized symbols shall be used and the handwriting shall be clear.

The above is stipulated in the *Electric Power Safety Working Regulations of State Grid Corporation of China*. For some local and municipal companies, the requirements are stricter, requiring that the work ticket is not allowed to be altered. If there are errors or omissions, it needs to be reprinted.

(2) Work tickets generated or printed by computer shall be in a uniform format, audited and verified by the issuer of the work order, and can only be executed after manual or electronic signature. A copy of the work ticket shall be kept at the workplace and collected by the person in charge of work. The other copy shall be collected by the work permitter and handed over per shift. The work permitter shall record the number, task, permit, and termination time of the work ticket in the register.

In normal circumstances, the work ticket cannot be implemented only after the issuer signs manually. The so-called "handover per shift" refers to the handover of work tickets according to the shift handover system.

(Ⅱ) Qualification of filling out and issuing

(1) For a work ticket, the work permitter and the person in charge of work shall not concurrently serve each other. If the issuer of the work ticket is also the work permitter or the person in charge of work, he/she shall have the corresponding qualifications and perform the corresponding safety responsibilities.

This article in the *Electric Power Safety Working Regulations of State Grid Corporation of China* was modified in 2014, from "The issuer of the work ticket, the person in charge of the work ticket, and the permitter of the work ticket shall not concurrently serve each other" into "The work permitter and the person in charge of work shall not concurrently serve each other".

(2) The work ticket shall be filled in by the person in charge of work, or by the issuer of the work ticket.

Generally speaking, the person in charge of work is responsible for filling in the work ticket. No matter the work ticket is filled in by the issuer or the person in charge, it can be effective after being checked and signed by the issuer.

(3) The work ticket shall be issued by the equipment O&M management organization (department), and may also be issued by the maintenance and capital construction organization that is qualified and approved by the equipment O&M management organization (department). The list of issuers of work tickets and persons in charge of work of the maintenance and capital construction organization shall be submitted in advance to the relevant equipment O&M management organization (dispatching control center) for the record.

(4) In contracted projects, work tickets can be issued in the form of "dual issuing". When issuing the work order, the issuer of the work ticket of both parties shall sign the work ticket respectively and bear the corresponding safety responsibility of the issuer of this part of the work ticket.

(5) When the power supply organization or the Construction Contractor carries out construction in the user's substation, the work ticket shall be issued by the Construction Contractor, Construction Contractor, or user that has the right to issue the work ticket.

(Ⅲ) Master/sub work ticket

(1) When there are more than two work sites listed in the first type of work ticket, or when there are two or more different work organizations (teams) working together, the master work ticket and sub work ticket can be used. The master work ticket and sub work ticket shall be issued by the same work ticket issuer. The safety measures listed on the master work ticket shall include all the safety measures listed on all sub work tickets. When several teams are working at the same time, only the person in charge of each sub work ticket is filled in the shift team member column of the master work ticket, and there is no need to fill in the names of all the staff in the shift team. The names of the staff in the shift team shall be filled out in the sub work ticket.

(2) The format of the master work ticket and sub work ticket is consistent with that of the first type of work ticket.

(3) The sub work ticket shall be made in duplicate, which shall be collected by the person in charge of the master work ticket and the person in charge of the sub work ticket. The approval and termination of the sub work ticket shall be handled by the person in charge of the master work ticket and the person in charge of the sub work ticket. The sub work ticket shall be approved only after the master work ticket is approved. The master work ticket shall not be terminated until all sub work tickets are terminated.

The issuer of the master work ticket and the sub work ticket is the same person. The person in charge of the master work ticket is the permitter of the sub work ticket, and the shift team members of the master work ticket are the persons in charge of the sub work tickets. During the implementation of sub work tickets, one of them is collected by the person in charge of the master work ticket and the other is collected by the person in charge of the sub work ticket.

Ⅲ. Use of work tickets

(1) A single person in charge of work cannot execute multiple work tickets at the same time. The work sites listed on the work ticket are limited to one electrical connection part.

① The so-called electrical connection part refers to the part of an electrical device that can be separated from other electrical devices using an isolating switch (knife switch).

② With DC bipolar disabled, the converter transformer and all high voltage DC equipment can be regarded as an electrical connection part.

③ With DC monopole operating, The disabled pole converter transformer, valve hall, DC equipment and water cooling system can be regarded as an electrical connection part. The bipolar public area is for operating equipment.

(2) The equipment under maintenance listed on a work ticket shall be turned off and turned on at the same time, and all safety measures specified in the work ticket shall be completed at one time. If part of the work has not been completed by the scheduled time, and the work needs to be continued without hindering the power transmission, the new work ticket shall be handled according to the electrification of the on-site equipment after the power transmission, and the safety measures shall be arranged before the work can be continued.

(3) If the following equipment is turned off and turned on simultaneously, the same work ticket can be used:

① Several electrical connection parts that belong to the same voltage class and are located in the same plane, and do not touch live conductors during work.

② A transformer is under interruption maintenance and its circuit breaker (switch) is also under interruption maintenance.

③ There is a power outage in the whole station.

(4) For the same type of work without power outage on several electrical connections in the same substation, a second type of work ticket can be used.

(5) A live-wire work ticket can be used for the same type of live-wire work in the same substation.

(6) For entering a substation or power plant step-up substation with a line or cable-related work ticket for work on overhead lines, cables, etc., the number of work tickets shall be added, approved by the substation or power plant work permitter, and retained. The list of issuers of work tickets and persons in charge of work of the above organizations shall be submitted in advance to the relevant O&M organization for the record.

(7) When there is a need to change the members of the shift team, the work can only be carried out with the consent of the person in charge of work and after the safe disclosure for the new

operators. The person in charge of work may not be changed except under special circumstances. If it is necessary to change the person in charge of work, the work ticket issuer shall agree and notify the work permitter, who will record the change on the work ticket. The person in charge of work is allowed to be changed once. The original and current persons in charge of work shall hand over and take over work tasks and safety measures.

After the change of the shift team members, the original "Shift team members" column in the work ticket will not be changed, but relevant content will be added in the "Staff changes" column. The new shift team members shall go through the disclosure procedures and sign in the column "Confirm the work tasks and safety measures assigned by the person in charge of work" before they can work.

The person in charge of work is the most important role in the on-site work, the change of person in charge of work will bring certain security risks, so it can not be changed except under special circumstances. If, after a change of the person in charge of work, the new person in charge cannot continue to be responsible for the on-site work due to force majeure, the work ticket must be terminated ahead of schedule, and other persons in charge shall go through the work ticket procedures again before they can resume their work.

(8) For adding work tasks within the scope of power outage and safety measures of the original work ticket, the person in charge of work shall obtain the consent of the issuer of the work ticket and the work permitter, and add the work items on the work ticket. If safety measures need to be changed or added, a new work ticket shall be filled out and the permit issuance procedures shall be completed again.

(9) For changing the person in charge of work or adding work tasks, if the work ticket issuer (and work permitter) cannot handle it in person, contact by telephone and mark on the work ticket register and work ticket.

(10) The first type of work ticket shall be delivered to the O&M personnel one day before work, and can be directly delivered or transmitted through fax or LAN. However, the work ticket permit sent by fax shall be fulfilled upon the arrival of the official work ticket. Temporary work can be handed over directly to the work permitter before the work begins. The second type of work ticket and live-wire work ticket can be handed over to the work permitter in advance on the day on which the work is carried out.

(11) When the work ticket is damaged and cannot be used again, a new work ticket shall be filled in and the permit issuance procedures shall be completed again.

Ⅳ. Period of validity and extension of work tickets

(1) The period of validity of the first and second types of work tickets and live-wire work tickets shall be limited to the approved maintenance period.

(2) If the first and second types of work tickets need to be extended, the person in charge of work shall apply to the person in charge of operation and maintenance before the end of the

construction period (for equipment under maintenance under the jurisdiction and permission of the control center, approval from the control personnel on duty is also required), the person in charge of operation and maintenance shall notify the work permitter to handle it. The first and second types of work tickets can only be extended once. The live-wire work tickets shall not be extended.

Ⅴ. Power transformation-related "five types of people"

(Ⅰ) Basic conditions for personnel listed on the work ticket

(1) An issuer of the work ticket shall be the production leader, technician, or person approved by the organization who is familiar with the technical level of personnel, equipment situation, these Regulations, and has relevant work experience. The list of issuers of work tickets shall be published.

(2) The person in charge of work (supervisor) shall have relevant work experience, be familiar with the equipment situation and these Regulations, and be approved by the work area (workshop, the same below). The person in charge of work shall also be familiar with the work abilities of the shift team members.

(3) The work permitters shall be O&M personnel or maintenance operators with certain work experience and approved by the work area (personnel who carry out the task operation and take safety measures); The work permitters for user substations and power distribution substations shall be high-voltage electrical workers with valid certificates.

(4) Special supervisors shall be persons with relevant work experience, familiar with the equipment situation and these Regulations.

(Ⅱ) Safety responsibilities of personnel listed on the work ticket

1. Work ticket issuer

(1) Confirm the necessity and safety of work.

(2) Confirm that the safety measures on the work ticket are correct and complete.

(3) Confirm the appropriateness and adequacy of the assigned persons in charge of work and shift team members.

2. Person in charge of work (supervisor)

(1) Organize work correctly and safely.

(2) Check whether the safety measures listed on the work ticket are correct and complete, and whether they meet the actual on-site conditions. If necessary, supplement and improve them.

(3) Before starting work, perform disclosure concerning the work tasks, safety measures, technical measures to the shift team members, and inform them of dangerous points, and confirm that each shift team member has signed.

(4) Strictly implement the safety measures listed on the work ticket.

(5) Supervise the shift team members to abide by these Regulations, correctly use labour protective equipment and safety tools and instruments, and implement on-site safety measures.

(6) Pay attention to the physical and mental health of the shift team members for any signs of abnormalities, and whether personnel changes are appropriate.

3. Work permitter

(1) Review whether the safety measures listed in the work ticket are correct, complete, and in compliance with the site conditions.

(2) Check whether the safety measures arranged at the work site are perfect or not. If necessary, supplement them.

(3) Be responsible for checking whether is any danger of sudden power connection to the equipment under maintenance.

(4) For the content listed in the work ticket, even if there is a little doubt about it, the work permitter shall consult the work ticket issuer and make it clear, and make a detailed supplement if necessary.

4. Special supervisor

(1) Define supervised personnel and the range of supervision.

(2) Explain the safety measures within the range of supervision to supervised personnel and inform them of dangerous points and safety precautions.

(3) Supervise the compliance with the Working Regulations and on-site safety measures by supervised personnel, and correct unsafe behaviors of supervised personnel in a timely manner.

5. Members of shift team

(1) Get familiar with the work content and work process, grasp safety measures, understand dangerous points during work, and sign for disclosure and confirmation on the work ticket.

(2) Obey the direction of the person in charge of work (supervisor) and special supervisor, strictly comply with the Working Regulations and labor disciplines, work within specified range of operation, be responsible for their own behaviors during work, and care about each other's work safety.

(3) Use construction tools and instruments, safety tools and instruments, and labour protective equipment correctly.

Among "five types of persons" engaged in power transformation, the work ticket issuer, the person in charge of work, and the work permitter are also call "three types of persons".

任务三　工作许可制度及其应用

一、许可流程

检修班组在电气设备上的任何工作，必须事先经过工作许可人同意，未办理许可手续，不得擅自进行工作。工作许可手续应通过一定的书面形式进行，发电厂、变电站通过工作票履行工作许可手续。

工作许可人在完成施工现场的安全措施后，还应完成以下手续，工作班方可开始工作：

（1）会同工作负责人到现场再次检查所做的安全措施，对具体的设备指明实际的隔离措施，证明检修设备确无电压。

（2）对工作负责人指明带电设备的位置和注意事项。

（3）和工作负责人在工作票上分别确认、签名。

工作票中所列“安全措施”均应由工作许可人及其运行班组完成，完成后，许可人会同负责人到现场，负责人检查所有安全措施是否与工作票“安全措施”项中所列一致；许可人负责告知负责人工作范围、停电设备与保留带电部位、现场安全措施布置，并进行现场验电，在负责人确认清楚后，双方在工作票“工作许可栏”签字并写下当前时间。

二、注意事项

运维人员不得变更有关检修设备的运行接线方式。工作负责人、工作许可人任何一方不得擅自变更安全措施，工作中如有特殊情况需要变更时，应先取得对方的同意并及时恢复。变更情况及时记录在值班日志内。

工作过程中的开关状态、接地状态、标示牌等安全措施，如果擅自改变，容易引起安全事故，故无特殊情况，不允许改变。有部分工作（如隔离开关整体调整等）需要分合接地开关，通常负责人在写安全措施时，“接地”部分会采取装设接地线的方式，而非合接地刀闸的方式，就避免了在工作过程中申请安全措施变更。

变电站（发电厂）第二种工作票可采取电话许可方式，但应录音，并各自做好记录。采取电话许可的工作票，工作所需安全措施可由工作人员自行布置，工作结束后应汇报工作许可人。

Task Ⅲ Work permit system and its application

Ⅰ. Permit process

Any work on electrical equipment carried out by the overhaul team shall obtain the consent of the work permitter. Work shall not be carried out without handling the permit procedure. The work permit procedure shall be handled in a certain written form. The power plant and substation shall perform the work permit procedure by means of the work ticket.

After completing safety measures on the construction site, the work permitter shall also handle the following procedures before the shift team starts working:

(1) Go to the construction site and check the safety measures again jointly with the person in charge of work, indicate actual isolation measures for specific equipment, and prove that there is no voltage on the equipment under maintenance.

(2) Indicate the position of live equipment and precautions to the person in charge of work.

(3) Confirm and sign on the work ticket together with the person in charge of work respectively.

All the safety measures listed in the work ticket shall be performed by the work permitter and its shift team. After performance, the permitter and the person in charge shall go to the site, and the person in charge shall check whether all the safety measures are consistent with the content listed in the item of safety measures in the work ticket; The permitter shall inform the person in charge of the work range, equipment with power failure, live parts and the arrangement of safety measures on site, and verify live parts on site. After the person in charge confirms clearly, they shall sign and write the current time on the work permit column in the work ticket.

Ⅱ. Precautions

O&M personnel shall not change the operation and wiring mode of equipment under maintenance. Neither the person in charge of work nor the work permitter shall change the safety measures without authorization. In case any change is required under special circumstances, one party shall obtain the consent of the other party and shall restore the safety measures timely. Changes shall be recorded in the log timely.

Changes in safety measures such as the switch state, earthing state, and sign board during work may easily cause safety accidents. Therefore, changes are not allowed without special circumstances. For some work (such as the overall adjustment of isolating switch), the grounding switch needs to be engaged or disengaged. Usually, when the person in charge is writing safety measures, a grounding wire will be installed on the grounding part instead of engaging the grounding knife-switch, which avoids the application for change in the safety measures during work.

The method of telephone permit may be adopted for the second type of work ticket for the substation (power plant). Sound recording and document record shall be made. For the work ticket which adopts telephone permit, the safety measures required for work may be arranged by the worker and a report shall be submitted to the work permitter after the work is over.

任务四　工作监护制度及其应用

工作许可手续完成后，工作负责人、专责监护人应向工作班成员交待工作内容、人员分工、带电部位和现场安全措施，进行危险点告知，并履行确认手续，工作班方可开始工作。工作负责人、专责监护人应始终在工作现场，对工作班人员的安全认真监护，及时纠正不安全的行为。

只有许可手续完成后，工作负责人才能带领工作班成员进入工作现场，然后召开“班前会”，告知所有工作班成员当前工作的内容、时间、地点，现场安全措施、危险点及预防措施、现场保留带电部位等信息，待所有成员确认清楚后，在工作票“确认工作负责人布置的工作任务和安全措施”栏目签字后，方可开始工作。

所有工作人员（包括工作负责人）不许单独进入、滞留在高压室、阀厅内和室外高压设备区内。若工作需要（如测量极性、回路导通试验、光纤回路检查等），而且现场设备允许时，可以准许工作班中有实际经验的一个人或几人同时在它室进行工作，但工作负责人应在事前将有关安全注意事项予以详尽的告知。

工作负责人在全部停电时，可以参加工作班工作。在部分停电时，只有在安全措施可靠，人员集中在一个工作地点，不致误碰有电部分的情况下，方能参加工作。

工作负责人、专责监护人应始终在工作现场。工作票签发人或工作负责人，应根据现场的安全条件、施工范围、工作需要等具体情况，增设专责监护人和确定被监护的人员。专责监护人不得兼做其他工作。专责监护人临时离开时，应通知被监护人员停止工作或离开工作现场，待专责监护人回来后方可恢复工作。若专责监护人必须长时间离开工作现场时，应由工作负责人变更专责监护人，履行变更手续，并告知全体被监护人员。

更变专责监护人应当在工作票“备注”栏写清楚原因、时间以及被变更的人员姓名。

工作期间，工作负责人若因故暂时离开工作现场时，应指定能胜任的人员临时代替，离开前应将工作现场交待清楚，并告知工作班成员。原工作负责人返回工作现场时，也应履行同样的交接手续。若工作负责人必须长时间离开工作现场时，应由原工作票签发人变更工作负责人，履行变更手续，并告知全体作业人员及工作许可人。原、现工作负责人应做好必要的交接。

若现场无他人能够胜任工作负责人，则负责人暂时离开时，应当告知工作班组暂时停止工作，待负责人回到工作现场时，方可继续工作。

Task Ⅳ Work supervision system and its application

After handling of the work permit procedure, the person in charge of work and the special supervisor shall explain the work content, division of labor, live parts and safety measures on site to members of the shift team, inform them of dangerous points, and go through the procedure for confirmation before the shift team starts to work. The person in charge of work and the special supervisor shall always stay on site to carefully supervise the safety of members of the shift team and timely correct unsafe behaviors.

The person in charge of work shall not bring members of the shift team to the work site until the permit procedure is completed. Then, the person in charge of work shall hold a pre-shift meeting and inform all the members of the shift team of the content, time and place of current work, as well as safety measures, dangerous points and preventive measures and live parts remained on site. After all the member confirm such information clearly and sign in the column "confirm the work task and safety measures assigned by the person in charge of work" in the work ticket.

Anyone of the workers (including the person in charge of work) shall not enter or stay in the high-voltage room, valve hall and outdoor high-voltage equipment area alone. In case certain work (such as polarity measurement, circuit conduction test, optical fiber circuit inspection, etc.) is required and the equipment on site permits, one or more members of the shift team who have practical experience are allowed to work in other rooms. However, the person in charge of work shall inform them of the detailed safety precautions in advance.

In case of power failure in all lines, the person in charge of work may participate in the work carried out by the shift team. In case of power failure in some lines, the person in charge of work shall not participate in the work until reliable safety measures are provided and all the personnel gather on the same work site and will not touch live parts accidentally.

The person in charge of work and the special supervisor shall always stay on the work site. The work ticket issuer or the person in charge of work shall designate an additional special supervisor and confirm the supervised personnel according to the safety conditions, range of construction and work requirements on site. The special supervisor shall not do other work at the same time. During the period when the special supervisor leaves the work site temporarily, the special supervisor shall inform the supervised personnel of stopping work or leaving the work site. The supervised personnel shall not resume work until the special supervisor comes back. In case the special supervisor must leave the work site for a long time, the person in charge of work shall change the special supervisor, go through the procedure for such change, and inform all the supervised personnel of such change.

For the change of special supervisor, the reason for and time of such change as well as the name of the person to be changed shall be stated clearly in the "Remarks" column of the work ticket.

During the work, if the person in charge of work needs to leave the work site temporarily for some reason, he/she shall assign a competent person to take over his/her work for the time being, explain matters on the work site clearly and inform members of the shift team before leaving. When the person in charge of work returns to the work site, he/she shall go through the same handover procedure. In case the person in charge of work must leave the work site for a long time, the previous work ticket issuer shall change the person in charge of work, go through the procedure for such change, and inform all the operators and the work permitter of such change. The previous person in charge of work shall make necessary handover to the current person in charge of work.

In case no person on site is competent to serve as the person in charge of work, before leaving the work site, the person in charge shall inform the shift team of stopping work for the time being. The shift team shall not resume work until the person in charge returns to the work site.

任务五　工作间断、转移和终结制度及其应用

一、工作间断

工作间断时，工作班人员应从工作现场撤出。每日收工，应清扫工作地点，开放已封闭的通道，并电话告知工作许可人。若工作间断后所有安全措施和接线方式保持不变，工作票可由工作负责人执存。次日复工时，工作负责人应电话告知工作许可人，并重新认真检查确认安全措施是否符合工作票要求。间断后继续工作，若无工作负责人或专责监护人带领，作业人员不得进入工作地点。

在未办理工作票终结手续以前，任何人员不准将停电设备合闸送电。在工作间断期间，若有紧急需要，运维人员可在工作票未交回的情况下合闸送电，但应先通知工作负责人，在得到工作班全体人员已经离开工作地点、可以送电的答复后方可执行，并应采取下列措施：

（1）拆除临时遮栏、接地线和标示牌，恢复常设遮栏，换挂“止步，高压危险!”的标示牌。

（2）应在所有道路派专人守候，以便告诉工作班人员“设备已经合闸送电，不得继续工作”。守候人员在工作票未交回以前，不得离开守候地点。

检修工作结束以前，若需将设备试加工作电压，应按下列条件进行：

（1）全体作业人员撤离工作地点。

（2）将该系统的所有工作票收回，拆除临时遮栏、接地线和标示牌，恢复常设遮栏。

（3）应在工作负责人和运维人员进行全面检查无误后，由运维人员进行加压试验。工作班若需继续工作时，应重新履行工作许可手续。

二、工作转移

在同一电气连接部分用同一张工作票依次在几个工作地点转移工作时，全部安全措施由运维人员在开工前一次做完，不需再办理转移手续。但工作负责人在转移工作地点时，应向作业人员交待带电范围、安全措施和注意事项。

三、工作终结

全部工作完毕后，工作班应清扫、整理现场。工作负责人应先周密地检查，待全体作业人员撤离工作地点后，再向运维人员交待所修项目、发现的问题、试验结果和存在问题等，并与运维人员共同检查设备状况、状态，有无遗留物件，是否清洁等，然后在工作票上填明工作结束时间。经双方签名后，表示工作终结。

待工作票上的临时遮栏已拆除，标示牌已取下，已恢复常设遮栏，未拆除的接地线、未拉开的接地刀闸（装置）等设备运行方式已汇报调控人员，工作票方告终结。

工作终结与工作票终结不同：工作终结手续应当是负责人带领班组人员将现场恢复至开始工作前的状态，然后负责人会同许可人（运维人员）进入现场，进行签字确认终结手续；而工作票终结则是许可人班组（运维班组）进入现场将现场恢复至送电前的状态，再通知调控人员，签字后方可终结工作票。

只有在同一停电系统的所有工作票都已终结，并得到值班调控人员或运维负责人的许可指令后，方可合闸送电。

已终结的工作票、事故紧急抢修单应保存 1 年。

Task Ⅴ The system of work interruption, transfer and completion and its application

Ⅰ. Work interruption

In case of work interruption, members of the shift team shall evacuate from the work site. After work is completed everyday, they shall clean up the work site, open closed channels, and inform the work permitter of such matters by telephone. If all the safety measures and wiring modes remain unchanged after work interruption, the work ticket may be kept by the person in charge of work. Upon resumption of work on the next day, the person in charge of work shall inform the work permitter by telephone, and carefully check whether the safety measures accord with the requirements of work ticket again. Upon resumption of work after interruption, operators shall not enter the work site without the leadership of the person in charge of work or the special supervisor.

Before handling of the procedure for completing the work ticket, anyone shall not close the knife switch and supply power to the equipment with power failure. In case of any emergency during work interruption, the O&M personnel shall first inform the person in charge of work, and get the reply that all the members of the shift team have left the work site and the supply of electricity is feasible. Then, the O&M personnel may close the knife switch and supply power before the work ticket is returned, and shall also take the following measures:

(1) Remove temporary barriers, grounding wires and sign boards, restore permanent barriers, and hang the sign board indicating "Stop! High voltage, danger!".

(2) Assign special personnel to watch on each road, in order to tell members of the shift team that "the knife switch has been closed and power has been supplied for equipment, and they shall not continue their work". All the watchmen shall not leave the site before the work ticket is returned.

Before maintenance is completed, if an operating voltage needs to be exerted on the equipment, the test shall meet the following requirements:

(1) All the operators shall evacuate from the work site.

(2) Withdraw all the work tickets of such system, remove temporary barriers, grounding wires and sign boards, and restore permanent barriers.

(3) After the person in charge of work and the O&M personnel conduct an overall inspection and confirm that there is no problem, the O&M personnel can exert an voltage for test. If the shift team needs to proceed with their work, they shall go through the work permit procedure again.

Ⅱ. Work transfer

In case the same work ticket is used to transfer work on the same electrically connected part between several work sites, all the safety measures shall be performed by the O&M personnel for once before commencement of work and handling of the transfer procedure is not required. However, before the person in charge of work transfers the work site, he/she shall explain the range of live parts, safety measures and precautions to the operators.

Ⅲ. Completion of work

After completion of all the work, the shift team shall clean up and reorganize the site. First, the person in charge of work shall conduct a careful inspection. After all the operators have evacuated from the work site, the person in charge of work shall explain the repair items, problems identified, test results and existing problems to the O&M personnel. Besides, the person in charge of work shall work together with the O&M personnel to check the status and condition of equipment, confirm whether any objects are left and whether the equipment is clean, and then write the time of work completion on the work ticket. Both parties shall sign on the work ticket to indicate that the work is completed.

The work ticket is not completed until the temporary barrier on the work ticket has been removed, the sign board has been taken down, and the permanent barrier has been restored, and the remaining grounding wire and closed grounding knife-switch (device) have been reported to the regulator.

The completion of work is different from the completion of work ticket. In terms of the procedure for completion of work, the person in charge shall lead members of the shift team to restore the site to the state before work. Then, the person in charge shall enter the site jointly with the permitter (O&M personnel) and sign to confirm the procedure for completion of work; in terms of the completion of work ticket, the shift team (O&M team) of the permitter shall enter the site and restore the site to the state before power is supplied, and inform the regulator, and sign to confirm the completion of work ticket.

The knife switch shall not be closed until all the work tickets of the same system with power failure have been completed and the permit command of the regulator on duty or the person in charge of operation and maintenance have been obtained.

The work ticket and emergency repair order that have been completed shall be kept for 1 year. [Extracted from 6.6.7 of the *Electric Power Safety Working Regulations (Power Transformation) of State Grid Corporation of China*]

任务六　电气设备现场勘察实训

一、作业任务

5 人一小组，一人扮演工作票签发人，一人扮演工作负责人，一人扮演工作许可人，两人扮演班组成员，以“110 kV 智能变电站”为工作环境，根据具体检修任务，工作负责人或签发人进行现场勘察，并按要求正确填写现场勘察报告。

二、引用标准及文件

（1）《国家电网公司电力安全工作规程（变电部分）》。

（2）《国家电网公司十八项电网重大反事故措施》。

三、作业条件

应在良好的天气进行现场勘察；勘察人员精神状态良好，熟悉现场设备的带电状况和勘察时的风险。

四、作业前准备

1. 现场设备勘察基本要求及条件

勘察现场设备情况，查阅相关技术资料，包括历史数据及相关规程。

2. 危险点及预防措施

（1）高压触电。

危险点：勘察设备均应视为带电设备。

预防措施：与设备保持足够的安全距离。

（2）使用不合格的安全工器具。

预防措施：巡视设备前，应正确选取安全工器具并检查合格。

3. 现场主接线图（图 3-2）

五、作业规范及要求

（1）现场勘察的基本情况：包括勘察的单位、班组、人员、时间、地点等信息。

（2）现场设备的双重名称与工作内容：应当在现场勘察记录报告中抄录现场设备的双重名称；详细写出现场工作的工作内容。

（3）现场设备或现场线路的具体情况：应在现场勘察记录报告中详细描述现场设备、线路目前的具体情况，包括但不仅限于线路的跨越、交叉、相关设备的尺寸、带电情况、相关距离等。必要时应当使用完善的图文相互配合的形式来描述。

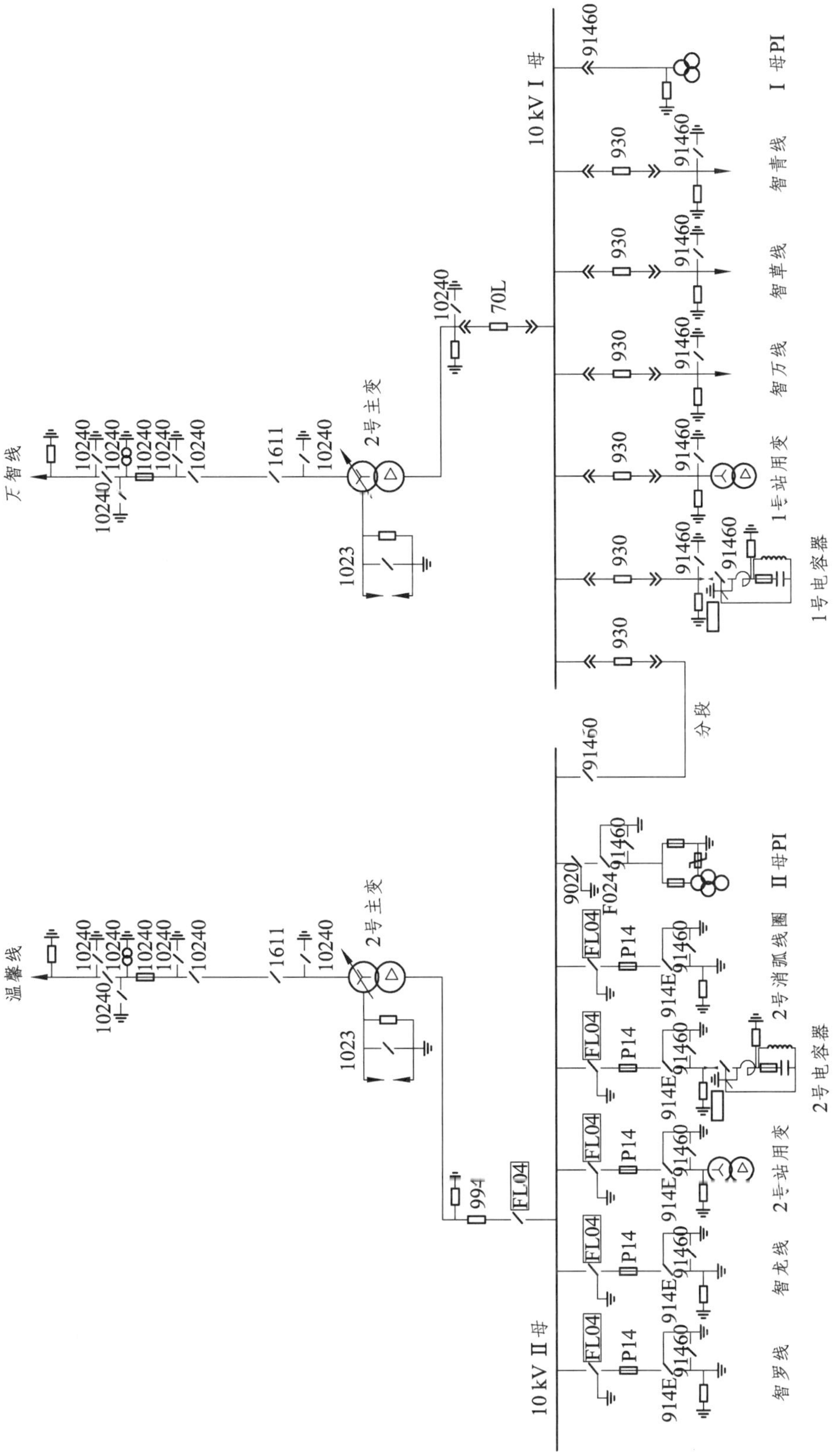

图 3-2　110 kV 实训智能变电站电气主接线图

（4）工作现场需要停电的部位与保留的带电部位：应在现场勘察记录报告中详细描述现场工作时现场带电的设备、工作现场会停电的设备以及停电设备上将保留的带电部位。

（5）工作现场的危险点：应在现场勘察记录报告中详细描述在现场工作时，可能发生的所有类型的危险点，包括但不仅限于高压触电、低压触电、高处坠落、物件伤人、油污等化学物质伤人、设备损伤等。

（6）工作现场应当采取的安全措施：根据上述第三条的工作中的危险点，应在现场勘察记录报告中详细描述工作前应当布置好的安全措施，包括但不仅限于应拉合的隔离开关、断路器，应装设的接地线，应悬挂的标示牌，应装设的围栏，等。

（7）现场工作时应当注意的事项：根据上述第四点的工作中的危险点以及安全措施，应在现场勘察记录报告中详细描述工作时应当注意的事项。

勘察报告应当以电子档形式上交，负责人撰写报告时，应当在现场拍照，并插入报告中加以阐述。

（8）现场勘察报告（表 3-3）。

表 3-3　现场勘察报告

现场勘察报告			
单位名称		班组名称	
地点		时间	
勘察负责人		记录人	
待检修设备双重名称			
现场工作内容			
现场设备具体情况			
现场应布置的安全措施			
现场危险点及预防措施			

六、作业流程及标准（表 3-4）

表 3-4　现场勘察流程及评分标准

班级		姓名		学号		考评员		成绩	

序号	作业名称	质量标准	分值/分	扣分标准	扣分	得分
1	工作准备					
1.1	着装穿戴	穿工作服、工作鞋，戴安全帽、线手套	10	未穿工作服、工作鞋，未戴安全帽、线手套，每缺少一项扣 1 分； 着装穿戴不规范，每处扣 1 分		
2	工具检查					
2.1	取安全帽和常用工器具，并检查	安全帽：检查外观、标签、合格证、下颌带是否完好； 检查电筒外观、标签、照明度	10	未取安全帽扣 2 分，未检查安全帽外观、标签、合格证、下颌带是否完好扣 1 分，未佩戴扣 5 分； 未检查电筒外观、标签、照明度扣 1 分，未拿手电筒扣 5 分		
3	勘察报告	现场勘察的基本情况：包括勘察的单位、班组、人员、时间、地点等信息；现场设备的双重名称与工作内容：应当在现场勘察记录报告中抄录现场设备的双重名称；详细写出现场工作的工作内容、现场设备或现场线路的具体情况：应在现场勘察记录报告中详细描述现场设备、线路目前的具体情况，包括但不仅限于线路的跨越、交叉、相关设备的尺寸、带电情况、相关距离等，必要时应当使用完善的图文相互配合的形式来描述。 工作现场需要停电的部位与保留的带电部位：应在现场勘察记录报告中详细描述现场带电的设备、工作现场会停电的设备以及停电设备上将保留的带电部位。	80	现场勘察的基本情况错、漏一项扣 2 分； 双重名称错误扣 10 分，工作内容错误扣 10 分； 报告中详细描述现场设备、线路目前的具体情况错、漏一项扣 2 分； 工作现场需要停电的部位与保留的带电部位错、漏一项扣 2 分； 工作现场的危险点错、漏一项扣 5 分； 工作现场应当采取的安全措施错、漏一项扣 5 分； 现场工作时应当注意的事项错、漏一项扣 2 分		

续表

序号	作业名称	质量标准	分值/分	扣分标准	扣分	得分
3	勘察报告	工作现场的危险点：应在现场勘察记录报告中详细描述在现场工作时，可能发生的所有类型的危险点，包括但不仅限于高压触电、低压触电、高处坠落、物件伤人、油污等化学物质伤人、设备损伤等。 工作现场应当采取的安全措施：根据上述第三条的工作中的危险点，应在现场勘察记录报告中详细描述工作前应当布置好的安全措施，包括但不仅限于应拉合的隔离开关、断路器、应装设的接地线、应悬挂的标示牌、应装设的围栏等。 现场工作时应当注意的事项：根据上述第三条的工作中的危险点以及安全措施，应在现场勘察报告中详细描述工作时应当注意的事项				
合计			100			

Task Ⅵ Training on site survey of electrical equipment

Ⅰ. Operating tasks

One team consists of 5 people. One serves as the work ticket issuer, one serves as the person in charge of work, one serves as the work permitter, and two serve as members of the shift team. Under the working environment of "110 kV intelligent substation", according to the specific overhaul task, the person in charge of work or the work ticket issuer shall conduct a site survey and fill in the site survey report correctly as required.

Ⅱ. Referenced standards and documents

(1) *Electric Power Safety Working Regulations (Power Transformation) of State Grid Corporation of China*;

(2) *18 Major Anti-accident Measures for Power Grid of State Grid Corporation of China.*

Ⅲ. Operating conditions

The site survey shall be carried out under a good weather condition. The surveyors shall be in good mental state and familiar with the state of electrification of equipment on site and the risk of survey.

Ⅳ. Preparation before operation

1. Basic requirements and conditions for on-site equipment survey

Conduct a survey of equipment on site and refer to relevant technical data, including historical data and related procedures.

2. Dangerous points and preventive measures

(1) High voltage electric shock.

Dangerous points: All the equipment under survey shall be regarded as live equipment.

Prevention and control measures: A sufficient safe distance shall be kept from the equipment.

(2) The use of unqualified safety tools and instruments

Prevention and control measures: Before routine inspection, the operator shall select correct safety tools and instruments and check whether they are qualified.

3. Main wiring diagram on site (see Fig. 3-2)

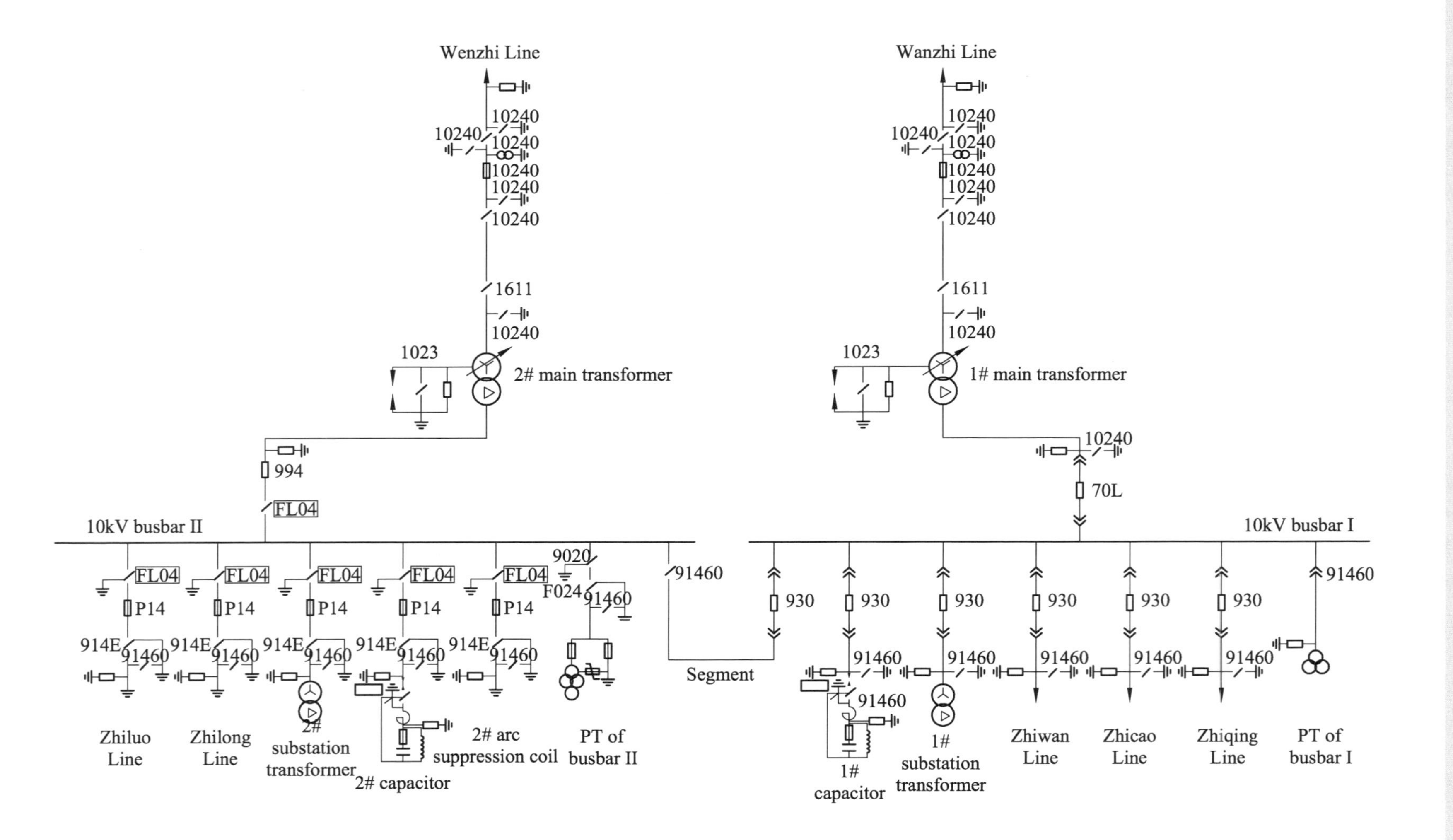

Fig. 3-2 Main electrical wiring diagram of 110 kV intelligent substation for training

V. Operating specifications and requirements

(1) Basic information of on-site survey: including the survey unit, team, members, time, place, etc.

(2) Double names and work content of equipment on site: The double names of equipment on site shall be copied in the report on site survey record; the details of work content on site shall be written.

(3) Specific conditions of equipment or lines on site: The current specific conditions of equipment and lines shall be described in detail in the report on site survey record, including but not limited to spanning and crossing of lines, dimensions, state of electrification and relative distance of equipment. If necessary, combine text with photos to describe related information.

(4) Parts for which interruption is required and live parts remained on the work site: Live equipment, equipment to which power failure will occur and live parts remained on the equipment with power failure on the work site shall be described in detail in the report on site survey record.

(5) Dangerous points on the work site: All kinds of potential dangerous points shall be described in detail in the report on site survey report, including but not limited to high voltage electric shock, low voltage electric shock, fall from height, personal injury caused by objects, personal injury caused by chemical substances such as oil pollutant, and equipment damage, etc.

(6) Safety measures required on the work site: According to the dangerous points described in the above Section III, safety measures which shall be provided shall be described in detail in the report on site survey record, including but not limited to isolating switches and circuit breakers to be closed, grounding wires to be installed, sign boards to be hung and fences to be installed.

(7) Precautions required on the work site: According to the dangerous points and safety measures described in above Section III, precautions required shall be described in detail in the report on site survey record.

The survey report shall be submitted in electronic form. During the preparation of the report, the person in charge shall take photos on site and insert them into the report for illustration.

(8) Site survey report (see Tab. 3-3).

Tab. 3-3 Site survey report

Site survey report			
Unit name		Team name	
Location		Time	
Person in charge of survey		Recorded by	
Double names of equipment under maintenance			

Continued

Work content on site	
Specific conditions of equipment on site	
Safety measures required on site	
Dangerous points and preventive measures on site	

Ⅵ. Operation procedures and standards (see Tab. 3-4)

Tab. 3-4 Site survey procedures and scoring standards

Class		Name		Student ID		Examiner		Score	
S/N	Operation name	Quality standard			Points	Deduction criteria		Deduction	Score
1	Work preparation								
1.1	Wearing	Wear work clothes and work shoes; wear safety helmet and cotton gloves			10	Deduct 1 point if failing to wear work clothes, work shoes, safety helmet or cotton gloves; Deduct 1 point if wearing does not meet the requirements			
2	Tool inspection								
2.1	Take safety helmet and common tools and instruments, and check them	Safety helmet: check the appearance, label, and certificate of conformity and confirm that the chin strap is in good condition; Check the appearance, label and illuminance of the flashlight			10	Deduct 2 points if failing to take the safety helmet; deduct 1 point if failing to check the appearance, label and certificate of conformity of the helmet and the integrity of chin strap; deduct 5 points if failing to wear the safety helmet; Deduct 1 point if failing to check the appearance, label and illuminance of the flashlight; Deduct 5 points if failing to take the flashlight			

Continued

S/N	Operation name	Quality standard	Points	Deduction criteria	Deduction	Score
3	Survey report	Basic information of site survey: including the survey unit, team, members, time, place, etc.; Double names and work content of equipment on site: The double names of equipment on site shall be copied in the report on site survey record; Write the details of work content on site; Specific conditions of equipment or lines on site: The current specific conditions of equipment and lines shall be described in detail in the report on site survey record, including but not limited to spanning and crossing of lines, dimensions, state of electrification and relative distance of equipment. If necessary, combine text with photos to describe related information;	80	For the basic information of site survey, deduct 2 points if there is any error or omission in one item; Deduct 10 points if there is any error in double names and deduct 10 points if there is any error in the work content; For the specific conditions of equipment and lines on site described in the report, deduct 2 points if there is any error or omission in one item; For parts on which power interruption is required and live parts remained on site, deduct 2 points if there is any error or omission in one item; For dangerous points on the work site, deduct 5 points if there is any error or omission in one item;		
		Parts for which power interruption is required and live parts remained on the work site: Live equipment, equipment for which power interruption is required and live parts remained on the equipment with power failure on the work site shall be described in detail in the report on site survey record. Dangerous points on the work site: All kinds of potential dangerous points shall be described in detail in the report on site survey report, including but not limited to high voltage electric shock, low voltage electric shock, fall from height, personal injury caused by objects, personal injury caused by chemical substances such		For safety measures to be taken on the work site, deduct 5 points if there is any error or omission in one item; For precautions required on the work site, deduct 2 points if there is any error or omission in one item.		

Continued

S/N	Operation name	Quality standard	Points	Deduction criteria	Deduction	Score
3	Survey report	as oil pollutant, and equipment damage, etc.; Safety measures required on the work site: According to the dangerous points described in the above Section III, safety measures which shall be provided shall be described in detail in the report on site survey record, including but not limited to isolating switches and circuit breakers to be closed, grounding wires to be installed, sign boards to be hung and fences to be installed; Precautions required on the work site: According to the dangerous points and safety measures described in above Section III, precautions required shall be described in detail in the survey report.				
Total			100			

任务七　变电站（发电厂）第一种工作票的填写实训

一、任务

5人一小组，一人扮演工作票签发人，一人扮演工作负责人，一人扮演工作许可人，两人扮演班组成员，以“110 kV 智能变电站”为工作环境，根据“实操项目”所要求的具体检修任务，以及“现场勘察报告”，工作负责人或工作票签发人按规范填写一张变电站（发电厂）第一种工作票第1～6项。

二、引用标准及文件

（1）《国家电网公司电力安全工作规程（变电部分）》。

（2）《国家电网公司十八项电网重大反事故措施》。

三、作业前准备

室内笔试，准备纸笔。

四、作业规范及要求

（1）单位填写自己系部、班级全称，班组填写小组名称，编号不填。

（2）负责人、班组成员、工作地点及设备双重名称据实书写。

（3）工作任务根据“实操项目”书写。

（4）计划工作时间根据教师安排书写。

（5）根据现场勘察报告与《国家电网公司电力安全工作规程（变电部分）》所要求的的组织措施、技术措施，正确填写第6项安全措施。

（6）变电站（发电厂）第一种工作票：

附件：　　　　　　　　　　　　　　　　　　　　　　　　　　工作票编号：

变电站（发电厂）第一种工作票

单位________________________________编号______________

1. 工作负责人（监护人）__________________班组______________

2. 工作班人员（不包括工作负责人）__________________共_______人

3. 工作的变、配电站名称及设备双重名称____________________________

4. 工作任务

工作地点及设备双重名称	工作内容

5. 计划工作时间：自______年___月___日___时___分至______年___月___日___时___分

6. 安全措施（必要时可附页绘图说明）

应拉断路器（开关）、隔离开关（刀闸）	已执行
应装接地线、应合接地刀闸（注明确实地点、名称及接地线编号）	已执行
应设遮栏、应挂标示牌及防止二次回路误碰等措施	已执行

已执行栏目及接地线编号由工作许可人填写。

工作地点保留带电部分或注意事项 （由工作 票签发人填写）	补充工作地点保留带电部分和安全措施 （由 工作许可人填写）

工作票签发人签名________签发日期____年__月__日__时__分

7. 收到工作票时间____年__月__日__时__分

运行值班人员签名________工作负责人签名________

8. 确认本工作票 1～7 项

工作负责人签名________工作许可人签名________

许可开始工作时间____年__月__日__时__分

9. 确认工作负责人布置的工作任务和安全措施

工作班组人员签名：________________________________

____

10. 工作负责人变动情况

原工作负责人________离去，变更________为工作负责人

工作票签发人____年__月__日__时__分

11. 工作人员变动情况（变动人员姓名、日期及时间）________

____

____

工作负责人签名________

12. 工作票延期

有效期延长到____年__月__日__时__分

工作负责人签名____年__月__日__时__分

工作许可人签名____年__月__日__时__分

13. 每日开工和收工时间（使用一天的工作票不必填写）

收工时间				工作负责人	工作许可人	开工时间				工作负责人	工作许可人
月	日	时	分			月	日	时	分		

14. 工作终结

全部工作于____年__月__日__时__分结束，设备及安全措施已恢复至开工前状态，工作人员已全部撤离，材料工具已清理完毕，工作已终结。

工作负责人签名________ 工作许可人签名________

15. 作票终结

临时遮栏、标示牌已拆除，常设遮栏已恢复。未拆除或未拉开的接地线编号________等共______组、接地刀闸（小车）共______副（台），已汇报调度值班员。

工作许可人签名________ ____年__月__日__时__分

16. 备注

（1）指定专责监护人________ 负责监护________

______________________________（地点及具体工作）

（2）其他事项______________________________

五、作业流程及标准（表3-5）

表3-5 工作票填写评分标准

班级		姓名		学号		考评员		成绩	

序号	作业名称	质量标准	分值/分	扣分标准	扣分	得分
1	单位	正确填写单位全称	3	不符合规定填写扣3分		
2	编号	正确填写编号	2	未填写编号扣2分； 填写编号不正确一处扣1分，扣完为止（时间、票种、编号）		
3	工作负责人	填写工作负责人	3	不符合规定填写扣3分		
4	班组	填写班组	3	未填写具体班组名称扣3分		
5	工作班人员	填写工作班人员	5	未填写工作班人员扣5分； 工作班人员少于或等于10人未填写所有人员姓名每处扣2分		
6	工作人数	填写工作人数	5	未填写工作人数扣5分；填写工作人数与现场工作人数不相符或工作人数含工作负责人的扣3分		
7	工作的变配电站名称及设备双重名称	填写工作的变配电站名称及设备双重名称	5	未填写工作的变配电站名称及设备双重名称扣5分； 未填写变、配电名称扣2分； 未填写电压等级扣3分		
8	工作任务	填写工作地点及设备双重名称； 填写工作内容	10	未填写工作地点及设备双名称扣10分； 填写工作地点及设备双名称不清楚、不准确、不具体每处扣2分； 未填写工作内容扣5分； 填写工作内容不清楚、不准确、不具体每处扣2分		
9	计划工作时间	填写计划工作时间	5	未填写计划工作时间扣5分； 填写计划工作时间不规范一处扣2.5分		

续表

序号	作业名称	质量标准	分值/分	扣分标准	扣分	得分
10	安全措施	填写应拉断路器（开关）、隔离开关（刀闸）；填写应装接地线、应合接地刀闸、应设遮栏、应挂标示牌及防止二次回路误碰等措施	45	未使用双重名称一处扣5分； 应断开的二次电源未断开，一处扣10分； 围栏装设不正确，扣10分； 标示牌装设不正确，一处扣5分		
11	工作地点保留带电部分或注意事项	根据工作地点保留带电部分或注意事项	5	未填写工作地点保留带电部分或注意事项一处扣5分； 填写工作地点保留带电部分或注意事项不正确一处扣2分		
12	工作票票面	工作票填写规范工整，关键字词不得涂改	9	重要文字出现错误或涂改设备双重编号、接地线组数编号、动词（如拉、合、装设等）、时间等扣9分；涂改填补超过3处，扣9分； 工作票涂改一处扣2分		
合计			100			

Task Ⅶ Training on filling of the first type of work ticket for substation (power plant)

Ⅰ. Task

One team consists of 5 people. One serves as the work ticket issuer, one serves as the person in charge of work, one serves as the work permitter, and two serve as members of the shift team. Under the working environment of "110 kV intelligent substation", according to the specific overhaul task required by "Practice Program 1" and the "site survey report", the person in charge of work or the work ticket issuer shall fill in item 1-6 in the first type of work ticket for substation (power plant).

Ⅱ. Referenced standards and documents

(1) *Electric Power Safety Working Regulations (Power Transformation) of State Grid Corporation of China*;

(2) *18 Major Anti-accident Measures for Power Grid of State Grid Corporation of China.*

Ⅲ. Preparation before operation

Prepare paper and pen for indoor written examination.

Ⅳ. Operating specifications and requirements

(1) The unit shall write the full name of department and class, and the shift team shall write the team name, but the number is not required.

(2) The double names of the person in charge, members of the shift team, workplace and equipment shall be written according to the facts.

(3) The work task shall be written according to "Practice Program 1".

(4) The planned working hours shall be written according to the teacher's arrangement.

(5) Item 6 "safety measures" shall be filled correctly according to the organization measures and technical measures specified in the survey report and the *Electric Power Safety Working Regulations (Power Transformation) of State Grid Corporation of China.*

(6) The first type of work ticket for substation (power plant):

Appendix: **Number of work ticket:**

The first type of work ticket for substation (power plant)

Unit______________________________________ Number___________________________

1. Person in charge of work (supervisor)________ Shift team_____________

2. Members of the shift team (excluding the person in charge of work) ______________________

people in total ____

3. Name of substation and power distribution substation and double names of equipment under operation____

4. Work task

Workplace and double names of equipment	Work content

5. Planned working time: from hour/minute, MM/DD/YYYY to hour/minute, MM/DD/YYYY

6. Safety measures (a sheet of drawing may be attached for illustration if necessary)

Circuit breaker (switch) and isolating switch (knife switch) which shall be disconnected	Executed
Grounding wire which shall be installed and grounding knife-switch which shall be closed (the actual place, name and the number of grounding wire shall be indicated)	Executed
Barriers shall be set up and sign boards shall be hung, in order to prevent accidental touch with the secondary circuit	Executed

The columns which have been executed and the numbers of grounding wires shall be filled in by the work permitter.

Live parts or precautions remained on the work site (to be filled in by the work ticket issuer)	Supplement the live parts remained on the work site and safety measures (to be filled in by the work permitter)

Signature of the work ticket issuer ________

Date of issue ____Hour____Minute___ Month____Day____Year

7. Receiving time of work ticket

________Hour____Minute____Month____Day____Year

Signature of on-duty operator________Signature of the person in charge of work________

8. Confirm Item 1-7 of this work ticket

Signature of the person in charge of work_______Signature of work permitter ______

Permitted time for starting work _______Hour____Minute ___Month___Day___Year

9. Confirm the work task and safety measures specified by the person in charge of work

Signature of members of the shift team:

____

10. Change of the person in charge of work

The previous person in charge of work______has left and _____will serve as the person in charge of work

Work ticket issuer __________Hour___Minute___Month___Day___Year

11. Change of the staff (name of the person changed, and date and time of change)

____

____

Signature of the person in charge of work ______________

12. Extension of work ticket

The period of validity will be extended to________Hour_____Minute_____Month _____Day_____Year

Signature of the person in charge of work _____Hour___Minute___Month___Day___ Year

Signature of the work permitter ________Hour___Minute___Month___Day___Year

13. Start time and end time everyday (the work ticket used only for one day does not need to be filled in)

End time				The person in charge of work	Work permitter	Start time				Work permitter	The person in charge of work
Month	Day	Hour	Minute			Month	Day	Hour	Minute		

14. End of work

All the work was completed at _____Hour___Minute___Month___Day___Year. The equipment and safety measures have been restored to the state before commencement of work, all the workers have been evacuated, materials and tools have been cleaned up, and the work has been completed.

Signature of the person in charge of work ____________

Signature of work permitter ____________

15. Work ticket termination

Temporary barriers and sign boards have been removed, and permanent barriers have been restored. There are ______ groups of grounding wire numbers ___________ which have not been removed or withdrawn, and there are ______ pairs (sets) of grounding knife-switches (trolleys) which have not been removed or withdrawn. Such information has been reported to the on-duty dispatcher.

Signature of work permitter _______Hour____Minute____Month____Day____Year

16. Remarks

(1) Designated special supervisor ________ Responsible for supervision _______

(Place and specific work)

(2) Other matters

V. Operation procedures and standards (see Tab. 3-5)

Tab. 3-5 Scoring standards for the filling of work ticket

Class		Name		Student ID		Examiner		Score	
S/N	Operation name	Quality standard		Points	Deduction criteria			Deduction	Score
1	Unit	Fill in the correct unit name		3	Deduct 3 points if failing to accord with the requirements				
2	Number	Fill in the correct number		2	Deduct 2 points if failing to fill in the number; Deduct 1 point if failing to fill in the number correctly until all the points are deducted (time, ticket type and number)				
3	The person in charge of work	Fill in the person in charge of work		3	Deduct 3 points if failing to accord with the requirements				
4	Shift team	Fill in the shift team		3	Deduct 3 points if failing to fill in the name of specific shift team				

Continued

S/N	Operation name	Quality standard	Points	Deduction criteria	Deduction	Score
5	Members of shift team	Fill in the members of shift team	5	Deduct 5 points if failing to fill in the members of shift team; Deduct 2 points if failing to fill in the name of one of ten or less members of the shift team		
6	Number of workers	Fill in the number of workers	5	Deduct 5 points if failing to fill in the number of workers; Deduct 3 points if the number of workers filled fail to accord with the actual number of workers on site or the person in charge of work is included in the number of workers		
7	Name of substation and power distribution substation and double names of equipment under operation	Name of substation and power distribution substation and double names of equipment filled	5	Deduct 5 points if failing to fill the name of substation and power distribution substation and double names of equipment under operation; Deduct 2 points if failing to fill in the name of substation and power distribution substation; Deduct 3 points if failing to fill in the voltage class		
8	Work task	Fill in the double names of workplace and equipment; Fill in the work content	10	Deduct 10 points if failing to fill in the double names of workplace and equipment; Deduct 2 points if failing to fill in clear, accurate and specific double names of each workplace and equipment; Deduct 5 points if failing to fill in the work content; Deduct 2 points if failing to fill in clear, accurate and specific double names of each workplace and equipment		

Continued

S/N	Operation name	Quality standard	Points	Deduction criteria	Deduction	Score
9	Planned working time	Fill in the planned working time	5	Deduct 5 points if failing to fill in the planned working time; Deduct 2.5 points if failing to fill in the planned working time according to the standard		
10	Safety measures	Fill in the circuit breaker (switch) and isolating switch (knife switch) which shall be disconnected; Fill in the grounding wire which shall be installed and the grounding knife-switch which shall be closed; Barriers shall be set up and sign boards shall be hung, in order to prevent accidental touch with the secondary circuit	45	Deduct 5 points if failing to use double names; Deduct 10 points if failing to disconnect the secondary power supply which shall have been disconnected; Deduct 10 points if failing to install the fence correctly; Deduct 5 points if failing to install the sign board		
11	Live parts or precautions remained on the work site	Live parts or precautions shall be remained according to the work site	5	Deduct 5 points if failing to fill in the live parts or precautions remained on the work site; Deduct 2 points if failing to fill in correct live parts or precautions remained on the work site		
12	Work ticket surface	The work ticket shall be filled in according to the standard and keywords shall not be altered	9	Deduct 9 points if there is any error or alteration in important text, including double names of equipment, number of group of grounding wire, verbs (such as pulling, closing and installing) and time; Deduct 9 points if there are more than 3 alterations; Deduct 2 points if there is one alteration on the work ticket		
Total			100			

任务八　变电站第一种工作票执行流程演练实训

一、作业任务

5人一小组，一人扮演工作票签发人，一人扮演工作负责人，一人扮演工作许可人，两人扮演班组成员，以“110 kV 智能变电站”为工作环境，根据“实操项目”所填写的工作票，按照《国家电网公司电力安全工作规程（变电部分）》中组织措施的要求，完成整个工作票签发、许可、使用、终结流程，最终完成整张工作票填写。

二、引用标准及文件

（1）《国家电网公司电力安全工作规程（变电部分）》。
（2）《国家电网公司十八项电网重大反事故措施》。

三、作业前准备

室内笔试，准备纸笔和“实操项目”所填工作票。

四、作业规范及要求

（1）本任务主要为角色扮演，5人分别扮演4种角色，在工作票签发、许可、使用、终结流程中，按照《国家电网公司电力安全工作规程（变电部分）》中组织措施的要求，通过工作票的检查、签发，现场工作许可，班前会、班后会、工作终结等流程，完成各角色工作任务，承担各角色责任。

（2）根据具体任务内容以及指导教师要求，决定任务是否包含工作间断、转移、延期，以及是否更换工作班成员、工作负责人等其他流程，其他流程均应在工作票票面上体现。

五、作业流程及标准（表3-6）

表3-6　工作票执行流程演练评分标准

<table>
<tr><td>班级</td><td></td><td>姓名</td><td></td><td>学号</td><td></td><td>考评员</td><td></td><td>成绩</td><td></td></tr>
<tr><td>序号</td><td>作业名称</td><td colspan="4">质量标准</td><td>分值/分</td><td colspan="2">扣分标准</td><td>扣分</td><td>得分</td></tr>
<tr><td>1</td><td>扮演</td><td colspan="4">5人正确扮演各自角色，完成从工作票许可至工作票终结所有流程</td><td>50</td><td colspan="2">任意一项工作不正确或遗漏扣3分，扣完为止</td><td></td><td></td></tr>
<tr><td>2</td><td colspan="10">补全工作票</td></tr>
<tr><td>2.1</td><td>时间</td><td colspan="4">正确填写收到工作票时间、许可开始工作时间、工作票延期时间、工作终结时间等</td><td>15</td><td colspan="2">任意一项时间填写错误扣3分，扣完为止</td><td></td><td></td></tr>
</table>

续表

序号	作业名称	质量标准	分值/分	扣分标准	扣分	得分
2.2	签名	根据各自角色完成签发人、负责人、许可人等签字	25	签字位置错误或漏签或代签扣5分，扣完为止		
2.3	其他	按具体情况要求正确填写工作班成员变动、每日开工、收工时间、工作票终结情况、备注等栏目	5	错填或漏填一项扣1分，扣完为止		
3	工作票票面	工作票填写规范工整，关键字词不得涂改	5	涂改填补超过3处，扣5分；工作票涂改一处扣2分		
合计			100			

Task Ⅷ Drill and training on the implementation procedure for the first type of work ticket for substation

Ⅰ. Operating tasks

One team consists of 5 people. One serves as the work ticket issuer, one serves as the person in charge of work, one serves as the work permitter, and two serve as members of the shift team. Under the working environment of "110 kV intelligent substation", according to the work ticket filled in "Practice Program 2" and the organization measures specified in the *Electric Power Safety Working Regulations (Power Transformation) of State Grid Corporation of China*, perform the issuance, permit, use and completion of the work ticket, and finally fill in the whole work ticket.

Ⅱ. Referenced standards and documents

(1) *Electric Power Safety Working Regulations (Power Transformation) of State Grid Corporation of China*;

(2) *18 Major Anti-accident Measures for Power Grid of State Grid Corporation of China.*

Ⅲ. Preparation before operation

Prepare paper and pen and the work ticket required for "Practice Program 4" for indoor written examination.

Ⅳ. Operating specifications and requirements

(1) This Task Is a role play in which 5 people play four different roles respectively. During the process of issuance, permit, use and completion of the work ticket, according to the organization measures specified in the *Electric Power Safety Working Regulations* (*Power Transformation*) *of State Grid Corporation of China*, they shall perform the duty and assume the responsibility for respective role through the inspection and issuance of work ticket and work permit on site, as well as meetings before and after shift and the completion of work.

(2) According to the content of specific task and requirements of the instructor, they shall determine other procedures, such as whether the Task Involves work interruption, transfer and extension, whether members of the shift team and the person in charge of work need to be changed. Other procedures shall be reflected on the work ticket.

Ⅴ. Operation procedures and standards (see Tab. 3-6)

Tab. 3-6 Scoring standards for the drill of the implementation process of work ticket

Class		Name		Student ID		Examiner		Score	
S/N	Operation name	Quality standard			Points	Deduction criteria		Deduction	Score
1	Role play	5 people shall play their own role respectively and complete all the procedures from the permit of work ticket to the completion of work ticket			50	Deduct 3 points if there is any error or omission in any item until all the points are deducted			
2	Fill in the rest of the work ticket								
2.1	Time	Fill in the correct time for receiving the work ticket, the permitted time to start work, the extended time of work ticket, and time for the completion of work			15	Deduct 3 points if there is any error in one of such time until all the points are deducted.			
2.2	Signature	The work ticket issuer, the person in charge and the permitter shall sign their name according to their respective role			25	Deduct 5 points if there is any error or omission in the signature or the work ticket is signed by other people, until all the points are deducted			
2.3	Miscellaneous	Fill in relevant columns according to the specific requirements, including change in members of the shift team, start time and end time each day, completion of the work ticket, and remarks			5	Deduct 1 point if there is any error or omission in one item until all the points are deducted			
3	Work ticket surface	The work ticket shall be filled in according to the standard and keywords shall not be altered			5	Deduct 5 points if there are more than 3 alterations; Deduct 2 points if there is one alteration on the work ticket			
Total					100				

模块四 安全技术措施的应用

根据《国家电网公司电力安全工作规程（变电部分）（线路部分）》的规定，保证 安全的技术措施包括停电、验电、接地、悬挂标示牌和装设遮栏（围栏）。

上述措施由运行人员或有权执行操作的人员执行。

学习目标

（1）掌握保证安全的技术措施的步骤和内容。

（2）能正确停电。

（3）能正确使用高压验电器进行高压验电。

（4）能正确装挂接地线。

（5）能正确悬挂标示牌和装设遮栏（围栏）。

Module Ⅳ Application of technical measures for safety

In accordance with the *Electric Power Safety Working Regulations (Power Transformation) (Line) of State Grid Corporation of China*, technical measures for safety include power failure, verification of live parts, grounding, hanging of sign boards, and installation of barriers (fences).

The above measures shall be implemented by operators or people who have the right to implement.

Learning objectives:

(1) Grasp the steps and content of technical measures for safety.

(2) Be able to cut off power correctly.

(3) Be able to use high-voltage electroscope for the verification of high-voltage live parts.

(4) Be able to install grounding wires.

(5) Be able to hang sign boards and install barriers (fences) correctly.

任务一 停 电

检修设备停电时，应把各方面的电源完全断开（任何运行中的星形接线设备的中性点，应视为带电设备）。不论是中性点直接接地还是中性点不接地的系统，正常运行时中性点都存在位移电压，系统发生故障时，电位会升高达到额定电压的10%以上，如不断开中性点，就有可能将电压引到检修设备上，发生危险。

断路器在停用状态操作电源是不断开的，如控制的回路发生二次混线、误碰、误操作等，会使其操动机构动作而自动合闸使设备带电。再者，断路器分闸时可能由于触头熔融、机构故障、位置指示器失灵等，造成未开断或不完全开断而位置指示器却在断开位置，形成虚断，也会使人出现误判断。断开隔离开关，一是做到一目了然，二是使设备与电源之间保持空气间隙，保持较高的绝缘强度。

手车开关应拉至检修位置，使各方面至少有一个明显的断开点，对于有些设备无法观察到明显断开点的（仅限 GIS 组合电器）至少应有两个及以上指示已同时发生对应变化，能反映设备运行状态的电气和机械等指示在开断位置才能作为判定明显断开点的依据。与停电设备有关的变压器和电压互感器，必须将设备各侧断开，防止向停电检修设备反送电。

禁止在只经断路器（开关）断开电源或只经换流器闭锁隔离电源的设备上工作。应拉开隔离开关（刀闸），手车开关应拉至试验或检修位置，应使各方面有一个明显的断开点，若无法观察到停电设备的断开点，应有能够反映设备运行状态的电气和机械等指示。与停电设备有关的变压器和电压互感器，应将设备各侧断开，防止向停电检修设备反送电。

检修设备和可能来电侧的断路器（开关）、隔离开关（刀闸）应断开控制电源和合闸电源，隔离开关（刀闸）操作把手应锁住，确保不会误送电。操作电源是对断路器和隔离开关的控制电源的统称。断路器和隔离开关断开后，如果不断开它们的控制电源和合闸电源，可能会因为多种原因如试验保护、遥控装置调试失当、误操作等而被突然合上，造成检修设备带电。因此为确保安全，一是要断开控制电源，二是断开后锁住操作机构，做到双保险。

对难以做到与电源完全断开的检修设备，可以拆除设备与电源之间的电气连接。工作地点应停电的设备如下：

（1）检修的设备。

（2）与工作人员工作中正常活动范围的距离小于表4-1规定的设备。

表4-1 工作人员工作中正常活动范围与设备带电部分的安全距离

电压等级/kV	安全距离/m	电压等级/kV	安全距离/m
10及以下（13.8）	0.35	1 000	9.50
20、35	0.60	±50及以下	1.50
63（66）、110	1.50	±400	6.70
220	3.00	±500	6.80
330	4.00	±660	9.00

续表

电压等级/kV	安全距离/m	电压等级/kV	安全距离/m
500	5.00	±800	10.10
750	8.00		

注：表中未列电压按高一挡电压等级的安全距离确定。750 kV 数据是按海拔 2 000 m 校正的，其他等级数据按海拔 1 000 m 校正。±400 kV 数据是按海拔 3 000 m 校正的；海拔 4 000 m 时，安全距离为 6.80 m。

（3）在 35 kV 及以下的设备处工作，安全距离虽大于表 4-1 规定的安全距离，但小于表 4-2 规定的安全距离，同时又无绝缘隔板、安全遮栏措施的设备。

表 4-2　设备不停电时的安全距离

电压等级/kV	安全距离/m	电压等级/kV	安全距离/m
10 及以下（13.8）	0.70	1 000	8.70
20、35	1.00	±50 及以下	1.50
63（66）、110	1.50	±400	5.90
220	3.00	±500	6.00
330	4.00	±660	8.40
500	5.00	±800	9.30
750	7.20		

（4）带电部分在工作人员后面、两侧、上下，且无可靠安全措施的设备。

（5）其他需要停电的设备。

Task Ⅰ Power failure

Before power failure in the equipment under maintenance, cut off the power supply in all aspects thoroughly (the neutral point of any running equipment with star connection shall be regarded as live equipment). Whether the neutral point is directly grounded or not grounded, there exists displacement voltage at the neutral point during normal operation. In case of system fault, the potential will rise to above 10% of the rated voltage. If the neutral point is not disconnected, the voltage may be introduced to the equipment under maintenance, causing danger.

When the circuit breaker is out of service, the operating power supply is not disconnected. In case of secondary mixed lines, accidental touch or misoperationin the circuit under control, the operating mechanism will be actuated and automatically closed, causing the equipment to be electrified. Besides, during the opening of circuit breaker, due to some reasons such as melted contact terminal, mechanism fault and fault in the position indicator, the circuit breaker may not be open or not be open completely, while the position indicator is still in the open position, which may cause false opening or misjudgment. Disconnecting the isolating switch can make the status absolutely clear and keep an air clearance between equipment and power supply, so as to keep high insulating strength.

The handcart switch shall be pulled to the overhaul position, so that there is at least one disconnecting point in each position. For some equipment, the disconnecting point cannot be observed clearly (only for GIS composite apparatus), at least two indicators have changed simultaneously. Only those electrical and mechanical indicators which can reflect the operating condition of equipment can be used as the basis for determining an obvious disconnecting point. For transformers and voltage transformers related with the equipment with interruption, each side of them shall be disconnected, in case they supply power back to the equipment under interruption maintenance.

Anyone shall not work on the equipment only when the power supply for circuit breaker (switch) is disconnected or when the power supply for current converter is blocked and isolated. The isolating switch (knife switch) shall be disconnected and the handcart switch shall be pulled to the test or maintenance position. There shall be one obvious disconnecting point in each circuit. If the disconnecting point of equipment under interruption cannot be observed, electrical and mechanical indicators which reflect the operating condition of equipment shall be provided. For transformers and voltage transformers related with the equipment with interruption, each side of them shall be disconnected, in case they supply power back to the equipment under interruption maintenance.

For the equipment under maintenance and the circuit breaker (switch) and isolating switch (knife switch) which may be powered on, the control power supply and closing power supply shall be disconnected, and the operating handle of the isolating switch (knife switch) shall be locked, in case power may be supplied accidentally. The power supply for circuit breaker and the power supply for isolating switch are collectively called operating power supply. After the circuit breaker and the isolating switch are disconnected, if their control power supply and closing power supply are not disconnected, they may be closed accidentally, and the equipment under maintenance may be electrified due to many reasons such as test protection, improper commissioning of remote control and misoperation. To ensure safety, the control power supply shall be disconnected and the operating mechanism shall be locked.

If it is difficult to completely disconnect the equipment under maintenance from the power supply, the electrical connection between equipment and power supply may be removed. Equipment of which the power supply shall be cut off on the work site is shown as follows:

(1) Equipment under overhaul.

(2) Equipment with a distance from the range of normal activity of workers less than the value specified in Tab. 4-1.

Tab. 4-1 Safe distance between the range of normal activity of workers and live parts on the equipment

Voltage class/kV	Safe distance/m	Voltage class/kV	Safe distance/m
10 and below (13.8)	0.35	1,000	9.50
20, 35	0.60	± 50 and below	1.50
63 (66) and 110	1.50	±400	6.70
220	3.00	±500	6.80
330	4.00	±660	9.00
500	5.00	±800	10.10
750	8.00		

Note: If no voltage class is listed in the corresponding column in the table, the safe distance for voltage class with a higher level shall apply. The data for 750 kV are corrected according to the altitude of 2,000 m and the data for other voltage classes are corrected according to the altitude of 1,000 m. The data for ±400 kV are corrected according to the altitude of 3,000 m; the safe distance is 6.80 m for an altitude of 4,000 m.

(3) Equipment which operates at a voltage of 35 kV and below, the safe distance is larger than the value specified in Tab. 4-1 but smaller than the value specified in Tab. 4-2, and no insulating board or safety barrier is provided.

(4) Equipment on which live parts are located behind, on both sides, above and below the worker, and for which no reliable and safe measures are provided.

(5) Other equipment for which power interruption is required.

Tab. 4-2 Safe distance for equipment without interruption

Voltage class/kV	Safe distance/m	Voltage class/kV	Safe distance/m
10 and below (13.8)	0.70	1,000	8.70
20, 35	1.00	± 50 and below	1.50
63 (66) and 110	1.50	±400	5.90

Continued

Voltage class/kV	Safe distance/m	Voltage class/kV	Safe distance/m
220	3.00	±500	6.00
330	4.00	±660	8.40
500	5.00	±800	9.30
750	7.20		

任务二　验　电

一、高压设备的验电方法

高压设备的验电应使用高压验电器，高压验电器的外观结构如图 4-1 所示。

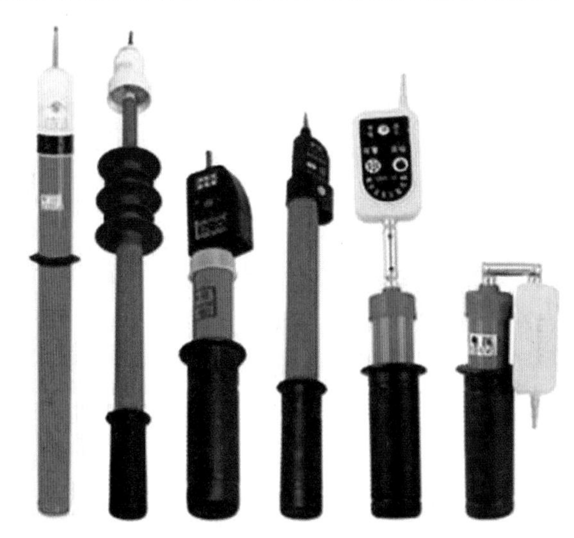

图 4-1　各种高压验电器

常见的高压验电器有回转式高压验电器和声光报警式验电器。

回转式高压验电器是利用带电导体尖端电晕放电产生的电晕风来驱动指示叶片旋转，从而检查设备或导体是否带电，也称风车式验电器，如图 4-2 所示。其结构主要由回转指示器和长度可以自由伸缩的绝缘棒组成。使用时将回转指示器触及线路或电气设备，若设备带电，指示叶片旋转，反之则不转。电压等级不同，回转式高压验电器配用的绝缘棒的节数及长度也不同，使用时，应选择合适的绝缘棒以保证测试人员的安全。这种验电器具有灵敏度高、选择性强、信号指示明确、操作方便等优点，不论在线路、杆塔上还是在变电所内部，都能够正确、明显地指示电力设备有无电压，广泛适用于 6 kV 及以上的交流系统。

声光报警式高压验电器是现阶段电力企业内最为常用的一种验电器，它用于检测 0.1 ~ 500 kV 线路或设备是否带有运行工频电压，以确保停电检修的工作人员的人身安全，其外观结构如图 4-3 所示。声光报警式高压验电器在验电时发出声和光双重报警信号，以提示工作人员被检线路或设备带电。声光报警式高压验电器具有很高的抗干扰性能、防短接能力、防电火花性能和辨别直流高压性能（对直流无反应，只是接触时短促响一声）。它具有自检测功能，只要一按自检测按钮，如果声光信号正常，即证明验电器指示器处于正常工作状态，可进行验电操作。

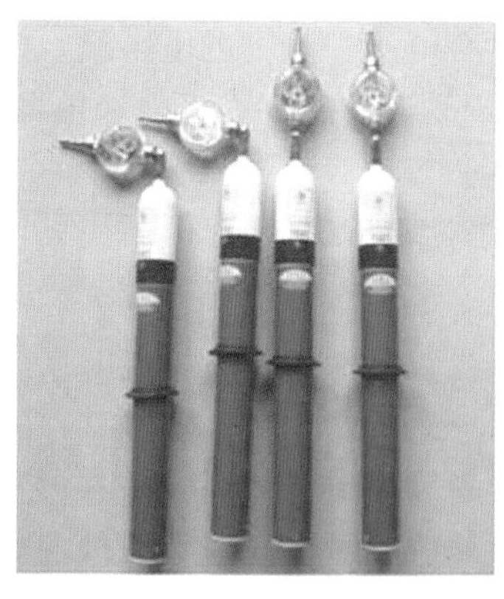

图 4-2　回转式高压验电器

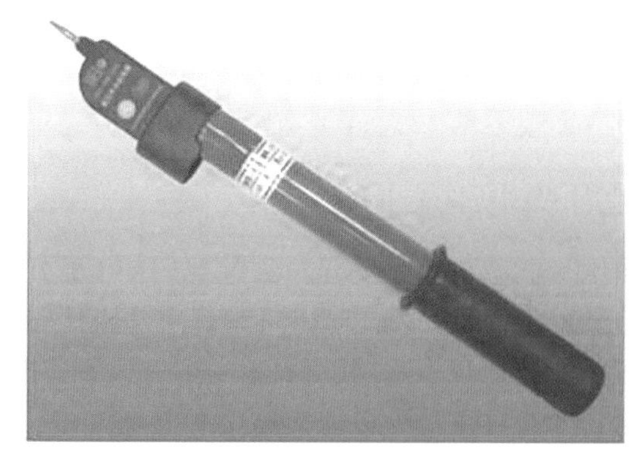

图 4-3 声光报警式高压验电器

高压验电前需要准备好安全工器具，主要包括：

（1）相应电压等级且合格的接触式验电器。验电是确认设备已无电压、防止发生带电挂地线（合接地刀闸）恶性误操作事故的有效手段。验电器应选取与所验设备电压等级相同的接触式验电器。如果验电器电压等级低于设备电压时，绝缘强度无法保证，操作人的人身安全将受到威胁；而高于设备电压时，可能达不到验电器发光、发声的启动电压，造成设备已无电压的误判断。

（2）验电时必须正确佩戴合格的安全帽，穿戴好经检查合格的绝缘手套和绝缘靴；将外衣袖口放入绝缘手套的伸长部分，裤管套入靴筒内，如图 4-4 所示。

图 4-4　高压验电

验电时，人体应与验电设备保持表 4-2 规定的安全距离，使用相应电压等级而且合格的接触式验电器，在装设接地线或合接地刀闸（装置）处对各相分别验电。验电前，应先在带电的设备上验电，证实验电器良好；注意验电器的工作触头不能直接接触带电体，只能逐渐接近带电体，直至验电器发出声、光或其他报警信号为止。无法在有电设备上进行试验时，可用工频高压发生器等确认验电器良好。验电应遵循先验离人体近的一相，由近及远、先低后高、先下后上等原则。不能只验一相，以防某一相仍然有电压，发生触电事故。然后在挂地线处进行三相分别验电，以防在某些意外情况下，可能出现其中一相带电而未被发现的情况。注意，验明无电的设备处应立刻挂装检验合格的三相接地线。

二、高压验电注意事项

（1）高压验电应戴绝缘手套。验电器的伸缩式绝缘棒长度应拉足，验电时手应握在手柄处，不得超过护环，雨雪天气时不得进行室外直接验电。

（2）对无法进行直接验电的设备、高压直流输电设备和雨雪天气时的户外设备，可以进行间接验电，即通过设备的机械指示位置、电气指示、带电显示装置、仪表及各种遥测、遥信等信号的变化来判断。判断时，至少应有两个非同样原理或非同源的指示发生对应变化，且所有指示均已同时发生对应变化，才能确认该设备已无电。以上检查项目应填写在操作票中作为检查项。检查中若发现其他任何信号有异常，均应停止操作，查明原因。若进行遥控操作，则应同时检查隔离开关（刀闸）的状态指示、遥测、遥信信号及带电显示装置的指示进行间接验电。330 kV 及以上的电气设备，可采用间接验电方法进行验电。

Task Ⅱ Verification of live parts

Ⅰ. Methods for the verification of live parts of high-voltage equipment

High-voltage electroscope shall be used for the verification of live parts of high-voltage equipment, and its appearance and structure are as shown in Fig. 4-1.

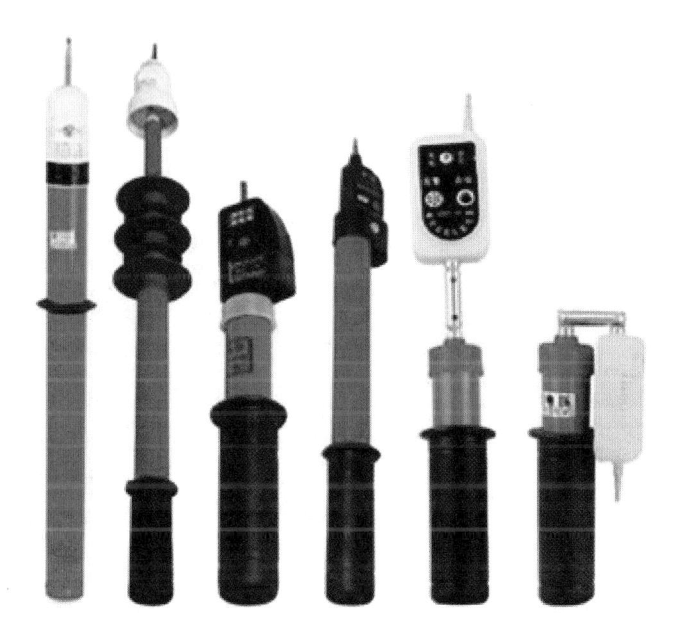

Fig. 4-1 Various high-voltage electroscopes

The commonly used high-voltage electroscope includes rotary high-voltage electroscope and audible and visual alarm type electroscope.

For rotary high-voltage electroscope, the corona wind generated by corona discharge at the tip of the live conductor is used to drive the indicator blade to rotate, so as to detect whether the equipment or conductor is electrified. It is also called windmill-type electroscope, as shown in Fig. 4-2. It is mainly composed of the rotary indicator and the insulating rod with flexible length. To use the electroscope, let the rotary indicator touch the line or electrical equipment. If the equipment is electrified, the indicator blade will rotate. Otherwise, the indicator blade will not rotate. The number of sections and the length of insulating rod on the high-voltage electroscope vary with voltage class. Suitable insulating rod shall be selected to ensure safety of the tester. Such electroscope has many advantages such as high sensitivity, high selectivity, clear signal indication, and easy operation. Whether it is installed on the line, tower or inside the substation, such electroscope is able to indicate whether there is any voltage on the electrical equipment accurately and clearly. It is widely used in AC systems with voltage class of 6 kV and above.

Audible and visual alarm type high-voltage electroscope is a type of electroscope most commonly used in power enterprises in the current stage. It is used to detect whether there is power frequency voltage in 0.1–500 kV line or equipment, in order to ensure the personal safety of interruption maintenance personnel. The appearance and structure are as shown in Fig. 4-3. The audible and visual alarm type high-voltage electroscope will send out audible and visual alarm signal, so as to give a prompt to the operator that the line or equipment under test is electrified. The audible and visual alarm type high-voltage electroscope is characterized by high anti-interference performance, the ability to prevent short circuit, electric spark and distinguish DC high voltage (with no response to DC voltage and only a short sound in case of touch). It has self-test function. Press the self-test button, if the audible and visual signal is normal, the electroscope indicator is working normally and can be used for verification of live parts.

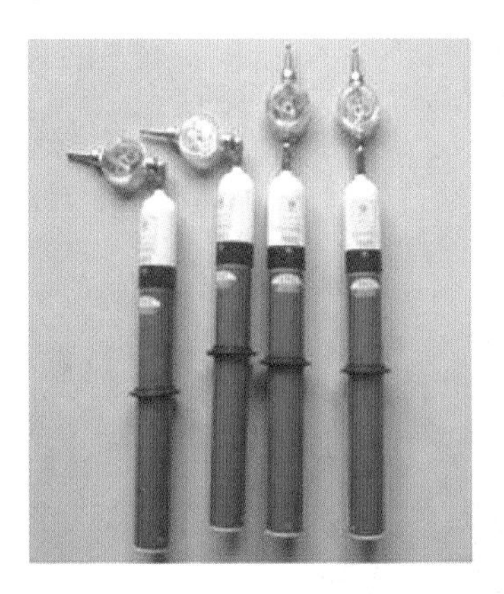

Fig. 4-2 Rotary high-voltage electroscope

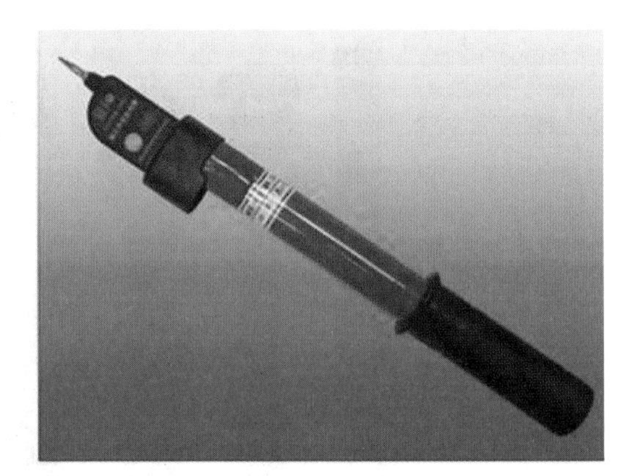

Fig. 4-3 Audible and visual alarm type high-voltage electroscope

Safety tools and instruments shall be prepared before verification of high-voltage live parts, mainly including:

(1) A qualified contact electroscope with corresponding voltage class. The verification of live parts is an effective means of confirming that there is no voltage on the equipment and prevent serious accidents caused by misoperation such as connecting the grounding wire when the power is on (closing the grounding knife switch). The contact electroscope with equal voltage class to that of the equipment under test shall be selected. If the voltage class of electroscope is lower than that of equipment, it cannot ensure insulating strength, which will cause a threat to the personal safety of the operator; if the voltage class of electroscope is higher than that of equipment, it cannot reach the starting voltage for the electroscope to generate an audible and visual alarm signal, resulting a misjudgment that there is no voltage on the equipment.

(2) The operator shall wear suitable and qualified safety helmet, as well as qualified insulating gloves and insulating boots during the verification of live parts; put the coat cuff into the extended part of insulating gloves and trouser leg into the boot leg, as shown in Fig. 4-4.

Fig. 4-4 Verification of high-voltage live parts

During the verification of live parts, the human body shall keep a safe distance from the equipment under verification, as specified in Tab. 4-2. A qualified contact electroscope with corresponding voltage class shall be used to verify each phase in the position where grounding wire or grounding knife switch (device) is installed. Before verification, conduct verification on live equipment to confirm that the electroscope is in good condition; the operating contact terminal of electroscope cannot touch live parts directly. Instead, it should gradually approach live parts until the electroscope generate an audible and visual or other alarm signal. In case you can not test live equipment, you can use a device such as a power frequency HV generator to check if the electroscope is in good condition. You should detect the phase that is closer to you. The general principle is to detect the electricity of equipment from far to near and from low to high. You can not test only one phase lest an electric shock accident should happen if there is still voltage in a phase. Then, you should detect three phases of the hanging grounding wire in case one of the phases is charged but not detected under some unexpected circumstances. Please note that the uncharged equipment shall be mounted right away with a three-phase grounding wire that is provenly compliant.

Ⅱ. Precautions for high-voltage detection

(1) Wear insulating gloves for high-voltage detection. The retractable insulating rod of the electroscope shall be extended long enough. When detecting electricity, you should hold the handle and not reach over the guard ring. Direct electricity detection is not allowed on rainy or snowy days.

(2) For equipment that can not be directly detected, HVDC transmission equipment, and outdoor equipment on rainy and snowy days, indirect power detection can be done, that is to say,

telling if there is electricity by the mechanical indicating position of the equipment, electrical indications, live display devices, instruments and a variety of changes in signals like telemetry signals and remote signals. You can only say that the equipment is not live only when at least two indications that are not of the same principle or source change accordingly and all indications change correspondingly at the same time. You should put down all the above items checked in the operation ticket as check items. In case any other signal is found abnormal during the checks, you should stop and find out the cause. When it comes to remote operation, you should detect electricity indirectly by checking the status indication of the isolating switch (knife switch), telemetry signals, remote signals, and the indication of the live display device at the same time. Electrical equipment of 330 kV and above can be detected by the indirect electricity detection method.

任务三 接 地

当验明设备确已无电压后，应立即装设经检验合格的三相接地线（图 4-5）并三相短路，如果是直流线路，则两级接地线分别直接接地。各工作班工作地段各端和有可能送电到停电线路工作地段的分支线（包括用户）都要验电。

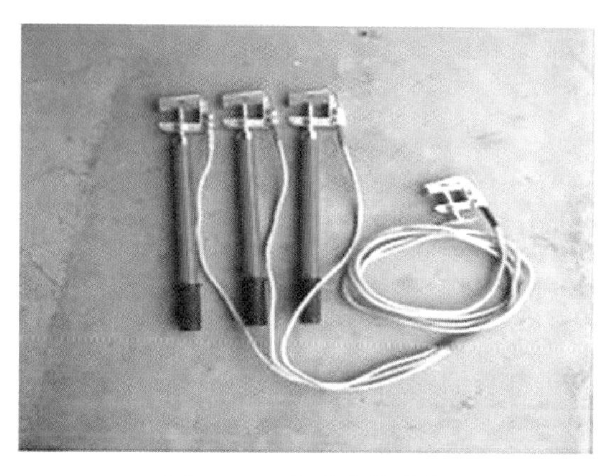

图 4-5 三相成套接地线

装设工作接地线：直流接地极线路，作业点两端应装设接地线。配合停电的线路可以只在工作地点附近装设一处工作接地线，如图 4-6 所示。

图 4-6 装设工作接地线

装设接地线是一项严肃谨慎的工作，若操作不当具有一定危险性。如果发生带电挂地线，不仅危及工作人员安全，而且可能会造成设备损坏。

装设接地线应由两人进行（经批准可以单人装设接地线的项目及运行人员除外），一人操作，一人监护，以确保装设地点、操作方法的正确性，防止因错挂、漏挂而发生误操作事故。两人同时进行工作，一旦发生带电挂地线，造成人身伤害事故时，还可起到相互救助的作用。

当验明设备确已无电压后，应立即将检修设备接地并三相短路。将检修设备三相短路接地，是保护工作人员在工作地点防止突然来电、避免伤害最可靠的措施。其作用是消除残存电荷和感应电压，使工作地点始终处于“地电位”的保护之中，确保人员不发生危险。如果工作中发生误送电，保护动作能使断路器跳闸，迅速切除电源，起到保护作用。装设三相短路接地线必须在验明设备确已无电压后立即进行，如果相隔时间较长，则应在装设前重新验电，这是考虑到在较长时间的间隔过程中，停电设备可能来电的意外情况。

电缆及电容器接地前应逐相充分放电，星形接线电容器的中性点应接地、串联电容器及与整组电容器脱离的电容器应逐个多次放电，装在绝缘支架上的电容器外壳也应放电。由于电缆及电容器属于容性设备，停电后短时间内仍然有大量剩余电荷，装设接地线时可能造成对操作人员的伤害，因此电缆及电容器在接地前应逐相、逐个多次放电，直至将剩余电荷放尽。由于星形接线的电容器，三相电容不平衡，中性点处会产生电压，工作时可能造成人身触电，因此电容器组中性点必须放电。串联电容器由于保险熔断、个别电容器与整组脱离等，造成电容器中的电荷可能没有完全放尽，所以串联电容器及与整组电容器脱离的电容器应逐个多次放电。装在绝缘支架上的电容器外壳也可能有感应电压，也应放电。

对于可能送电至停电设备的各方面都应装设接地线或合上接地刀闸（装置），所装接地线与带电部分应考虑接地线摆动时仍符合安全距离的规定。这样做是为了保证工作人员始终在接地线的保护范围内，防止检修设备突然来电和感应电压对人身造成伤害。对有可能产生感应电压的设备也应视为电源设备，应视情况适当增加接地线。在装设接地线时，要充分考虑接地线在遇大风摆动时与带电部分的安全距离，避免因接地线摆动导致运行设备接地故障。

对于因平行或邻近带电设备导致检修设备可能产生感应电压的情况，应加装工作接地线或使用个人保安线，如图 4-7 所示，加装的接地线应登录在工作票上，个人保安线由工作人员自装自拆。

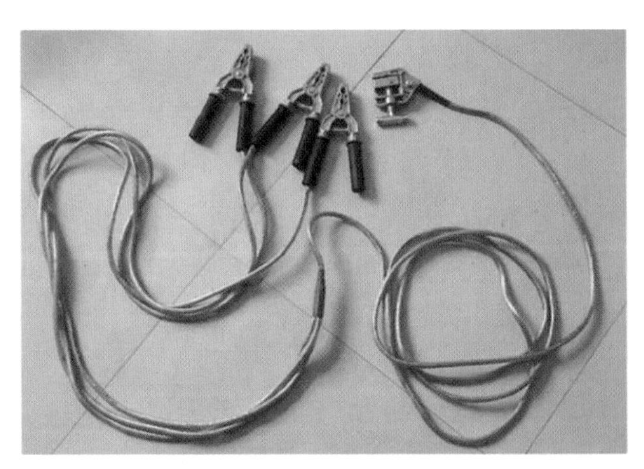

图 4-7　个人保安线

在门型构架的线路侧进行停电检修，如工作地点与所装接地线的距离小于 10 m，工作地点虽在接地线外侧，也可不另装接地线。

检修部分若分为几个在电气上不相连接的部分，如分段母线以隔离开关（刀闸）或断路器（开关）隔开分成几段，则各段应分别验电接地短路。降压变电站全部停电时，应将各个可能来电侧的部分接地短路，其余部分不必每段都装设接地线或合上接地刀闸（装置）。

接地线、接地刀闸与检修设备之间不得连有断路器（开关）或熔断器。若由于设备原因，接地刀闸与检修设备之间连有断路器（开关），在接地刀闸和断路器（开关）合上后，应有保证断路器（开关）不会分闸的措施。

在配电装置上，接地线应装在该装置导电部分的规定地点，这些地点的油漆应刮去，并画有黑色标记。所有配电装置的适当地点，均应设有与接地网相连的接地端，接地电阻应合格。接地线应采用三相短路式接地线，若使用分相式接地线时，应设置三相合一的接地端。

装设接地线应先接接地端，后接导体端，接地线应接触良好，连接应可靠。

拆接地线的顺序与此相反。装、拆接地线均应使用绝缘棒和戴绝缘手套。人体不得碰触接地线或未接地的导线，以防止触电。先装接地端后接导体端是操作安全的需要，在装、拆接地线的过程中，应始终保证接地线处于良好的接地状态，这样当设备突然来电时，能有效地限制接地线上的电位，保证装、拆地线的人员的安全。操作第一步即应将接地线的接地端与地极螺栓做可靠的连接，这样在发生各种故障的情况下都能有效地限制地线上的电位，然后再接到导体端；拆接地线时，只有在导体端与设备全部解开后，才可拆除接地端子上的接地线。否则，若先行拆除了接地端，则泄放感应电荷的通路被隔断，操作人员再接触检修设备或地线，就有触电的危险。由于在装拆接地线的过程中有感应电存在或突然来电的可能，操作人员必须戴好绝缘手套、穿绝缘鞋，使用绝缘棒。

未接地的导线有感应电或剩余电荷存在，地线在操作中也可能会有剩余电荷流过而带电，加之如果绝缘老化或破损，人体接触将会发生触电，因此使用绝缘棒装设时应注意选择好位置，避免人体与周围已停电设备或地线直接碰触。装设接地线还应注意使所装接地线与带电设备导体之间保持规定的安全距离。带接地线拆设备接头时，应采取防止接地线脱落的措施。成套接地线应用有透明护套的多股软铜线组成，其截面面积不得小于 25 mm^2，同时应满足装设地点短路电流的要求。禁止使用其他导线作接地线或短路线，接地线应使用专用的线夹固定在导体上，禁止用缠绕的方法进行接地或短路。接地线的选择有以下几方面要求：

（1）截面面积。接地线的作用是保持工作设备的地电位，截面积必须满足短路电流的要求，且最小不得小于 25 mm^2（按铜材料要求）。要使接地线通过高达数十千安的短路电流，而且该短路电流所产生的电压降不大于规定的安全电压值，就必须使它的阻抗值很小，故应选择截面积足够大的导电性能良好的金属材料线。

（2）满足短路电流热容量的要求。发生短路时，通过短路接地线作用于断路器使其跳闸。在断路器未跳闸或拒绝动作时，短路电流在接地线中产生的热量应不至于将它熔断。否则，工作地点将失去保护而使事故扩大。

（3）接地线必须具有足够的柔韧性和机械耐拉强度，耐磨，不易锈蚀。电力生产和电力工程上都选用带透明绝缘护套的多股软铜线来制作接地线。这是因为带透明绝缘护套的多股软铜线柔软不易折断，操作携带方便，导电性能好。软铜线外应包有透明的绝缘塑料（透明护套），既能保护地线免受磨损，又便于观察地线有无损坏。禁止使用其他导线作接地线或短路线，以防止使用普通导线不能满足动、热稳定要求，而不能起到保护工作人员安全的目的。接地线如果接触不良，在流过短路电流时，就会产生较大电压降加到检修设备上，造成人身伤害。如果接触电阻过大，在短路电流作用下甚至会造成接地线接地处烧断，因此装设接地线时必须使用完好的专用线夹固定在导体上，而接地线部分应固定在与接地网可靠连接的专用接地螺丝上或用专用的线夹固定在接地体上，并保证其接触良好。专用线夹应满足在短路电流作用下的动、热稳定要求。

禁止工作人员擅自移动或拆除接地线。高压回路上必须要拆除全部或一部分接地线后才可进行的工作，包括测量母线和电缆的绝缘电阻、测量线路参数、检查断路器（开关）触头是否同时接触等，如：拆除一相接地线；拆除接地线，保留短路线；接地线全部拆除或拉开接地刀闸（装置）。

上述工作应征得运行人员的许可（根据调度员指令装设的接地线，应征得调度员的许可），方可进行，工作完毕后立即恢复。接地线是保证工作人员生命安全的有效措施，擅自拆除或移动接地线将使工作地点失去地线保护，擅自移动接地线还有造成带电挂地线的危险，因此严禁工作人员擅自移动或拆除接地线。因工作需要必须将接地线拆除时，必须经值班员许可，采取相应的安全措施后，检修人员方可进行工作。工作完毕后应立即恢复原接地措施，防止遗漏接地线，缩短失去地线保护的时间。每组接地线均应编号，并存放在固定地点。存放位置亦应编号，接地线号码与存放位置号码应一致。接地线应有专门架柜存放，按组编号，使用完毕应检查整理对号入座。在管理方面应建立安全工具记录卡，由专人保管，负责维修，使其经常保持在完好的备用状态。接地线以组编号时按照不重复的原则编号。接地线存放位置也应按照同样组数固定排序，对号存放。这样便于掌握地线的使用和拆除情况，便于核查核对，便于加强管理，防止带地线合闸发生事故。

装、拆接地线，应做好记录，交接班时应交代清楚。对于已装设的接地线应用专门的登记簿将接地情况记录在案，并在交接班日志专门栏目中登记，以便于当班运行人员在送电前知晓开关接地情况，防止各种原因造成的带地线合闸。

Task Ⅲ Grounding

After confirming that the equipment is not applied with voltage, you should mount a compliant three-phase grounding wire (Fig. 4-5) and short the three phases. In the case of a DC line, the two-stage grounding wire shall be directly grounded. Each terminal of the shift team's work site and the branch lines (including users) that may supply power to the work site of the power-interrupted line shall be detected.

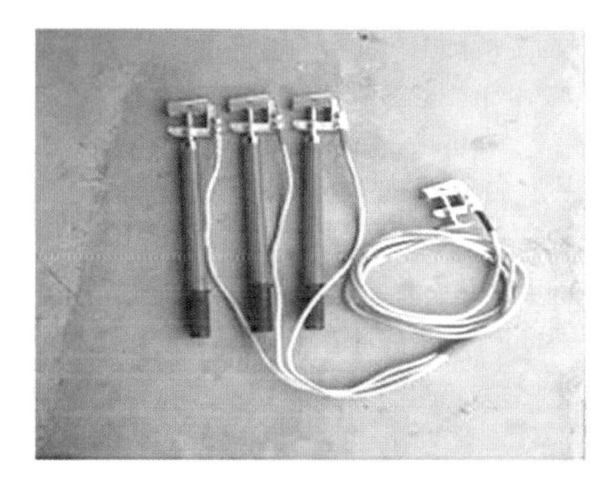

Fig. 4-5 Three-phase grounding wire set

Mounting working grounding wire. For a DC grounding wire, both ends of the operation point shall be mounted with grounding wires. For a line whose power is interrupted, you can mount a working grounding wire only around the work site, as shown in Fig. 4-6.

Fig. 4-6 Mounting working grounding wire

Mounting a grounding wire is serious and prudent work in that any misoperation would pose a certain risk. Any hanging grounding wire, if live, could endanger workers and damage equipment.

Two people are required to mount a grounding wire (except for projects which a single person is approved to mount the grounding wire and operators who get such approval), with one mounting and the other supervising, to ensure that where and how the grounding wire is mounted is correct and to prevent misoperation due to incorrect hanging or omission of any grounding wires. Two people working at the same time can help each other when a live electric grounding wire injures people.

After confirming that the equipment is voltage-free, the equipment should be grounded under maintenance immediately and short its three phases. Three-phase short-circuit grounding of the equipment under maintenance is the most reliable measure to protect the worker from a sudden resumption of power supply and avoid injury at the work site. It can eliminate residual charge and induced voltage so that the work site is always under the protection of "ground potential" and people are kept from danger. If undesired power transmission happens at work, the protection will trigger the circuit breaker to trip, quickly cutting off the power supply and protecting you and the equipment. The three-phase short-circuit grounding wires should be mounted right after confirming that the equipment is voltage-free. If there is a long interval between the confirmation and mounting, electricity should be re-detected before mounting. This is because the power-interrupted equipment may become live in this long interval.

Cables and capacitors shall be discharged phase by phase before grounding, the neutral point of star connection capacitors shall be grounded, the capacitors connected in series and capacitors separated from the whole bank of capacitors shall be discharged multiple times, one by one, and the capacitor housings mounted on the insulating supports shall also be discharged. Cables and capacitors are capacitive apparatus that may still retain a large amount of residual charge for a short period of time after the power interruption and thus may harm operators who are mounting the grounding wires. Therefore, cables and capacitors shall be discharged multiple times, phase by phase, before grounding, until the residual charge is discharged. At work, people may get electric shocks from a star connection capacitor with unbalanced three-phase capacitance because it will generate a voltage at the neutral point. Therefore, the neutral point of a capacitor bank must be discharged. Because a capacitor's fuse is blown, an individual capacitor is separated from the whole bank, or for some other reason, the capacitor may not be completely discharged. So, the capacitors connected in series and the capacitor separated from the whole bank shall be discharged multiple times, once a time. The capacitor housing mounted on the insulating support may also have an induced voltage and should be discharged as well.

Anything that could supply power to the power-interrupted equipment shall be connected to a grounding wire or the grounding knife-switch (device) shall be closed. The grounding wire and the live part shall meet the requirement for maintaining a safe distance when the grounding wire is swinging. In this way, workers will always be protected by the grounding wire, avoiding personal injuries resulting from the induced voltage of the equipment under maintenance when the power is back on suddenly. Equipment that may generate induced voltage should also be regarded as power supply equipment and connected to more grounding wires as appropriate. When mounting a grounding wire, its safe distance should be fully taken into account from the live part because a gale will swing it. This is to avoid any grounding faults of the operating equipment caused by the swing of the grounding wire.

The equipment under maintenance may generate induced voltage when parallel to or close to live equipment. To address this problem, a working grounding wire should be mounted or a personal safety grounding wire should be applied, as shown in Fig. 4-7. The installed grounding wire shall be recorded on the work ticket, and the personal safety grounding wire shall be mounted and removed on your own.

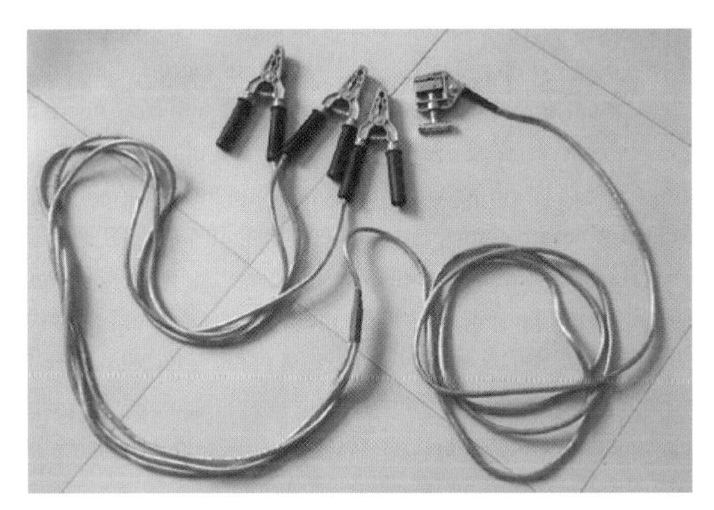

Fig. 4-7 Personal safety grounding wire

For interruption maintenance at the line side of the gantry structure, if the distance between the work site and the installed grounding wire is less than 10 m, additional grounding wires may not be required, though the work site is located somewhere away from the grounding wire.

When the part to be maintained is divided into several sections not electrically connected, such as a sectionalized busbar that is divided by isolating switches (knife switches) or circuit breakers (switches), you should detect, ground, and short these sections. When the power of all step-down substations is cut, parts on the side where the power is going to back on should be grounded or shorted, and for other parts, it's not necessary to mount grounding wires or close the grounding knife-switches (devices) for each section.

A circuit breaker (switch) fuse must not be connected between a grounding wire, grounding knife-switch, and equipment under maintenance. If a circuit breaker (switch) is connected between the grounding knife-switch and the equipment under maintenance due to the latter, measures should be taken to ensure that the circuit breaker (switch) will not open after the grounding knife-switch and circuit breaker (switch) close.

On a power distribution unit, the grounding wire shall be mounted at the specified point on the conducting part of the unit, where the paint shall be scraped off and marked in black. All power distribution units must be provided with a grounding terminal at an appropriate place, which is connected to the grounding grid, and the grounding resistance must be compliant. The grounding wires shall be three-phase short-circuit grounding wires. If a shunt-type grounding wire is used, a three-phase-in-one grounding terminal shall be provided.

When mounting a grounding wire, the grounding terminal should be connected first and then the conductor terminal. The grounding wire shall be in good contact with others and connected reliably.

The sequence to remove a grounding wire is the opposite of mounting it. When mounting and removing grounding wires, an insulating rod and wear insulating gloves should be worn. You should keep your body away from any grounding wires or ungrounded conductors to prevent electric shocks. For the sake of operational safety, the grounding terminal should be connected before the conductor terminal. When mounting and removing a grounding wire, make sure that the grounding wire is always well grounded so that its potential will be effectively limited when the power supply to the equipment is resumed, safeguarding the operator mounting or removing the grounding wire. The first step is to reliably connect the grounding terminal of the grounding wire to the grounding electrode bolt before connecting to the conductor terminal so that the potential of the grounding wire can be effectively limited in the event of any faults. When removing the grounding wire, the grounding wire should be remoued from the grounding terminal only after the conductor terminal is disconnected from the equipment completely. If the grounding terminal is removed first, the path to release induced charge is cut off, and then the operator may risk getting an electric shock when working on the equipment under maintenance or the grounding wire. Operators must wear insulating gloves, insulating shoes and use an insulating rod when mounting and removing grounding wires because of the possibility of induced charge presence or sudden resumption of power supply.

Ungrounded conductors may carry induced or residual charge, and grounding wires may remain live when residual charge flows through during operation. Besides, if the insulation is aging or damaged, people will get an electric shock from contact. So, when using an insulating rod to mount a grounding wire, a proper place should be chosen to avoid directly contacting the power-interrupted equipment nearby or the grounding wire. Maintaining the specified safe distance between the grounding wire to be mounted and the live equipment's conductor is crucial. When removing the connector of equipment with a grounding wire, measures should be taken to prevent the grounding wire from coming off. A complete grounding wire set shall be composed of multiple strands of annealed copper wires with transparent sheath, whose cross section shall not be less than 25 mm^2, and shall meet the requirements of short circuit current at the installation site. It's forbidden to use other conductors as grounding wires or short-circuit wires. The grounding wires shall be secured to the conductors with special wire clamps, and it's forbidden to wind grounding wires or short-circuit wires. The following requirements shall be met when selecting grounding wires:

(1) Crossing sectional area. The grounding wires are used to maintain the ground potential of the working equipment. The cross sectional area of a grounding wire must meet the requirements of the short circuit current and must be equal to or greater than 25 mm^2 (requirements for copper grounding wires). If tens of thousands of amps of short circuit current is to flow through a grounding wire and the voltage drop generated by the short circuit current is not greater than the specified voltage value, the grounding wire must have a very small impedance. So, metal wires with

good electrical conductivity and a large cross sectional area should be selected.

(2) Meeting the requirements of the thermal capacity of short circuit current. When a short circuit occurs, the short-circuit grounding wire triggers the circuit breaker to trip. If the circuit breaker does not trip or refuses to act, the heat generated by the short circuit current in the grounding wire should not fuse it. Otherwise, the protection of the work site will fail, escalating the accident.

(3) The grounding wire shall boast sufficient flexibility, mechanical tensile strength, and wear resistance, and shall not be prone to rusting. Multistranded annealed copper wires with a transparent insulating sheath are used to make grounding wires in power generation and power engineering. This is because multistranded annealed copper wires with a transparent insulating sheath are soft and not easily breakable, easy to work on and carry, and have good electrical conductivity. The annealed copper wires should be encased in transparent insulating plastic (transparent sheath) so that the grounding wires are protected from wear and you can observe if they are damaged. It is prohibited to use other conductors as grounding wires or short-circuit wires lest ordinary wires should not protect the worker because they cannot meet the requirements of dynamic stability and thermal stability. If the grounding wire is in poor contact, a large voltage drop will be generated in the equipment under maintenance when the short circuit current flows through, causing personal injury. If the contact resistance is excessive, where the grounding wire is grounded will be burnt under the action of the short circuit current. So, when mounting a grounding wire, it must be secured to a conductor with an intact special wire clamp, and the grounding wire shall be fixed on the special grounding screw reliably connected with the grounding grid or fixed on the grounding body with a special wire clamp, and ensure that the contact is good. The special wire clamp shall meet the requirements of dynamic stability and thermal stability under the action of the short circuit current.

Workers are forbidden to move or remove any grounding wires without permission. When it comes to working on an HV circuit, all or some of the grounding wires must be removed before you proceed to do the work like measuring the insulation resistance of a busbar or a cable, measuring the line parameters, and checking if the circuit breaker (switch) contact terminals are in contact at the same time. For example, remove the grounding wire of a phase; remove the grounding wires and retain the short-circuit wires; remove all grounding wires or open the grounding knife-switch (knife switch).

The above work shall be carried out only with the permission of the operator (if the grounding wire is mounted as commanded by the dispatcher, the dispatcher's permission shall be obtained).The grounding wires are an effective measure to ensure the safety of workers. Any unauthorized removal or movement of grounding wires will result in loss of protection of the work site, and unauthorized movement of the hanging ground wires may also make them live. So, it is strictly prohibited for workers to move or remove the grounding wires without authorization. When the grounding wires must be removed as required by work, the maintainer must ask the man on duty for permission and take corresponding safety measures before proceeding. After the work is done, the grounding measures shall be restored immediately, and be sure not to omit the grounding wires to shorten the time of losing protection. Each group of grounding wires shall be numbered and kept in a fixed place. The storage positions should also be numbered, and the grounding wire number should be the same as the storage location number. The grounding wires should be stored in a special rack cabinet and numbered by groups. After use, they shall be checked, organized, and returned to their original position. In terms of management, a safety tool record card shall be established, which is kept by a specially designated person who is responsible for maintenance to keep the grounding wires often in a good standby state. Groups of grounding wires shall not be numbered repeatedly. The storage positions of the grounding wires should also be numbered in accordance with the same wire group numbers, and the grounding wires shall be stored in the storage positions numbered accordingly. In this way, you can readily know which grounding wires are used and which are removed, perform checks and strengthen management, and prevent accidents caused by closing in the presence of ground wires.

When mounting and removing grounding wires, it should be recorded and be clearly informed the people who will work on the next shift. Mounted grounding wires shall be recorded on a special register and put down in the special column of the shift log so that the on-duty operator will if the switch is grounded before supplying power and prevent closing in the presence of grounding wires caused by various reasons.

任务四　悬挂标示牌和装设遮栏（围栏）

在一经合闸即可送电到工作地点的断路器（开关）和隔离开关（刀闸）的操作把手上，均应悬挂“禁止合闸，有人工作”的标示牌。

如果线路上有人工作，应在线路断路器（开关）和隔离开关（刀闸）操作把手上悬挂“禁止合闸，线路有人工作”的标示牌。

对由于设备原因，接地刀闸与检修设备之间连有断路器（开关），在接地刀闸和断路器（开关）合上后，在断路器（开关）操作把手上，应悬挂“禁止分闸”的标示牌。

在显示屏上进行操作的断路器（开关）和隔离开关（刀闸）的操作处均应相应设置“禁止合闸，有人工作”或“禁止合闸，线路有人工作”以及“禁止分闸”的标记。悬挂标示牌可提醒有关人员及时纠正可能进行的错误操作和行为，防止误向有人工作的设备（线路）合闸送电和误入误触带电部分。因此，在一经合闸即可送电到工作地点的断路器和隔离开关的操作把手上，均应悬挂“禁止合闸，有人工作”的标示牌。对同时能进行远方和就地操作的隔离开关，则还应在隔离开关操作把手上悬挂标示牌。

目前由于部分变电站断路器、隔离开关的操作在监控机上进行，所以在监控屏上进行操作的断路器和隔离开关的操作处均应相应设置“禁止合闸，有人工作”或“禁止合闸，线路有人工作”的标记。当线路有人工作时，则应在线路断路器和隔离开关的操作把手上悬挂“禁止合闸，线路有人工作”的标示牌，以提醒值班人员线路上有人工作，防止向有人工作的线路合闸送电。此标示牌的悬挂和拆除应按调度命令执行。

部分停电的工作，安全距离小于表 4-2 规定距离以内的未停电设备，应装设临时遮栏，如图 4-8 所示。临时遮栏与带电部分的距离不得小于表 4-1 的规定数值，临时遮栏可用干燥木材、橡胶或其他坚韧绝缘材料制成，装设应牢固，并悬挂“止步，高压危险!”的标示牌。

图 4-8　遮栏（围栏）

35 kV 及以下设备的临时遮栏，如因工作特殊需要，可用绝缘隔板与带电部分直接接触。

在室内高压设备上工作，应在工作地点两旁及对面运行设备间隔的遮栏（围栏）上和禁止通行的过道遮栏（围栏）上悬挂“止步，高压危险!”的标示牌。高压开关柜内手车开关拉出后，隔离带电部位的挡板封闭后禁止开启，并设置“止步，高压危险!”的标示牌。由于室内高压设备间隔紧凑，工作间隔的检修设备与两旁间隔、对面间隔带电设备相邻，一旦检修人员疏忽大意，没有仔细核对调度编号，很容易发生误入带电间隔的情况。所以为了防止检修人员误入带电间隔，在室内高压设备上工作时，应在工作地点的两旁间隔和对面间隔的遮栏（围栏）上悬挂“止步，高压危险!”的标示牌。某些通道，由于与带电设备或高压试验设备的距离不能满足安全距离规定，所以应在此类通道处设置遮栏（围栏）并悬挂“止步，高压危险!”的标示牌，以警戒他人不许通过。

在室外高压设备上工作，应在工作地点四周装设围栏，其出入口要围至邻近道路旁边，并设有“从此进出”的标示牌。工作地点四周围栏上悬挂适当数量的“止步，高压危险!”标示牌，标示牌应朝向围栏里面。若室外配电装置的大部分设备停电，只有个别地点保留有带电设备而其他设备无触及带电导体的可能时，可以在带电设备四周装设全封闭围栏，围栏上悬挂适当数量的“止步，高压危险!”标示牌，标示牌应朝向围栏外面。室外设备大都没有固定的围栏，设备布置也不像室内那样集中，工作地点人员多、范围大，往往有登高作业，监护工作困难，这就更有必要在工作地点装设围栏，限制作业人员的活动范围。围栏应采用封闭式网状遮栏，并具有独立支柱，在围栏四周面向围栏内悬挂适当数量的“止步，高压危险!”的标示牌，以警示检修人员只能在围栏内进行工作。为方便工作人员进出，围栏出入口要围至邻近道路旁边，并设有“从此进出”的标示牌。

为了减轻运行人员布置安全措施的工作量，同时又能满足工作安全需要，当室外大部分设备停电而其他设备无触及带电导体的可能时，只有个别地点保留带电，可将带电设备四周装设全封闭围栏。此时危险源在围栏内，为防止人员误入带电间隔，所以应在围栏上面向围栏外悬挂“止步，高压危险!”标示牌，禁止越过围栏。在工作地点设置“在此工作”的标示牌。室外为指明工作地点，将检修设备与运行设备加以明确的区分，在检修设备处设置“在此工作”的标示牌。一张工作票若有几个工作地点，均应设置“在此工作”标示牌；在交直流屏、保护屏、自动化屏等屏柜处工作时，应在屏柜前后分别设置“在此工作”标示牌。

在室外构架上工作，应在工作地点邻近带电部分的横梁上，悬挂“止步，高压危险!”的标示牌。在工作人员上下铁架或梯子上，应悬挂“从此上下”的标示牌。在邻近其他可能误登的带电构架上，应悬挂“禁止攀登，高压危险!”的标示牌。由于室外架构横梁往往与邻近间隔相连，且在高处作业时邻近带电设备不便区分，存在误登带电设备架构的隐患，所以在室外构架上工作，工作地点邻近带电部分的横梁上应面向工作架构悬挂“止步，高压危险!”的标示牌。运行人员一般不具备专门登高工具和熟练的登高技能，因此在办理工作许可手续时，由检修人员在工作许可人监护悬挂，在办理工作终结手续时仍在工作许可人监护下取下。为防止工作人员误登带电架构，运行人员应在工作人员上、下铁架或梯子上悬挂“从此上下”的标示牌。为了警示工作人员不要攀登到带电设备架构上，在邻近其他可能误登的带电架构上，悬挂“禁止攀登，高压危险!”的标示牌。

禁止工作人员擅自移动或拆除遮栏（围栏）、标示牌。因工作原因必须短时移动或拆除遮栏（围栏）、标示牌时，应征得工作许可人同意，并在工作负责人的监护下进行，完毕后应立即恢复。临时遮栏、接地线、标示牌、围栏等都是为保证检修工作人员的人身安全和设备的安全运行所做的措施，是使工作地点与带电间隔隔开、防止工作人员误碰带电设备，擅自变更安全措施，就可能变更工作范围，造成人身触电的危险。工作人员如因工作需要必须短时变动安全措施时，应征得工作许可人的同意，并在工作负责人监护下变更安全措施，在完成工作后，应立即恢复原来状态并报告工作许可人。若扩大工作任务需变更或增设安全措施者，必须填用新的工作票，并重新履行工作许可手续。直流换流站单极停电工作，应在双极公共区域设备与停电区域之间设置围栏，在围栏面向停电设备及运行阀厅门口悬挂“止步，高压危险!”标示牌。在检修阀厅和直流场设备处设置“在此工作”的标示牌。所悬挂的标示牌如图 4-9 所示。

图 4-9 标示牌

Task Ⅳ Hanging sign boards and mounting barriers (fences)

A sign board, reading "No switch closing! Working!", shall be hung on the operating handle of the circuit breaker (switch) and isolating switch (knife switch) that will supply power to the work site if close.

When someone is working on a line, a sign board, reading "No switch closing! Working on the line!", shall be hung on the operating handle of the circuit breaker (switch) and isolating switch (knife switch) of that line.

If a circuit breaker (switch) is connected between the grounding knife-switch and the equipment under maintenance due to the latter, a "No switch opening" sign board shall be hung on the operating handle of the circuit breaker (switch) after the grounding knife-switch and the circuit breaker (switch) close.

For the circuit breakers (switches) and isolating switches (knife switches) that are operated on the display screen, corresponding markers, reading "No switch closing! Working!", or "No switch closing! Working on the line!", and "No switch opening" shall be provided where the display screen is. Hanging sign boards can remind the relevant personnel to stop any possible mal-operations and improper behaviors in time and prevent falsely closing and transmitting power to the equipment (line) that someone is working on and mistakenly entering and touching any live parts. Therefore, a sign board, reading "No switch closing! Working!", shall be hung on the operating handle of the circuit breaker and isolating switch that will supply power to the work site if close. For an isolating switch that can be operated remotely and locally, a sign board shall also be hung on its operating handle.

At present, the circuit breakers and isolating switches of some substations are operated on the monitors. So, for the circuit breakers and isolating switches that are operated on the display screen, corresponding markers, reading "No switch closing! Working!", or "No switch closing! Working on the line!" shall be provided where the display screens are. When someone is working on a line, a sign board, reading "No switch closing! Working on the line!", shall be hung on the operating handle of the circuit breaker and isolating switch of that line to alert the operator on duty to someone working on the line so that he/she won't supply power to that line by closing. The sign board shall be hung and removed as instructed by the dispatcher.

When it comes to some work that requires power interruption, temporary barriers shall be provided for the non-power-interrupted equipment whose safe distance is smaller than the distance specified in Tab. 4-2, as shown in Fig. 4-8. The distance between the temporary barriers and the live part must not be smaller than the values specified in Tab. 4-1. Temporary barriers can be made of dry wood, rubber, or other tough insulating materials and shall be firmly mounted and a sign board reading "Stop! High voltage, danger!" shall be hung on them.

Fig. 4-8 Barrier (fence)

As to temporary barriers of equipment of 35 kV or below, If the special needs of the work, an insulating partition can be used to contact the live part directly if it is a special need of work.

When working on indoor HV equipment, a sign board reading “Stop! High voltage, danger!” should be hung on the barriers (fences) for the side compartment and opposite compartment of the operating equipment on the work site, as well as on the barriers (fences) of passages that are forbidden to pass. After the handcart switch in the HV switch cabinet is pulled out, the baffle plate used to isolate the live part, once closed, shall not be opened, and a sign board reading “Stop! High voltage, danger!” shall be provided. The indoor HV equipment compartments are compactly arranged. The equipment under maintenance in the working compartment is adjacent to live equipment in the side compartment and opposite compartment. If the maintainer acts recklessly and fails to check the dispatching number carefully, it's very likely that he/she will enter the electrified compartment by mistake. To prevent any maintainer from entering the electrified compartment by mistake, a sign board reading “Stop! High voltage, danger!” should be hung on the barriers (fences) for the side compartment and opposite compartment on the work site when the maintainer when it comes to working on indoor HV equipment. Because the distance between some passages and live equipment or HV test equipment does not meet the safety distance requirements, barriers (fences) shall be provided at such passages and a sign board reading “Stop! High voltage, danger!” shall be hung on the barriers warn others off the passages.

When working on outdoor HV equipment, a fence should be installed around the work site, with its entrance arranged next to an adjacent road, and a sign board reading “Enter and exit from here” shall be provided. An appropriate number of sign boards reading “Stop! High voltage, danger!” should be hung on the fences surrounding the work site, and the sign boards shall face what's inside the fences. If most of the equipment of the outdoor power distribution unit is power-interrupted, live equipment only exists in a few places, and other equipment is not likely to come in contact with the live conductors, a fully enclosed fence can be installed around the live equipment, an appropriate number of sign boards reading “Stop! High voltage, danger!” should be hung on the fence, and the sign boards shall face what's inside the fence. Most outdoor equipment has no fixed fences and is not laid out as concentrated as indoor equipment. The work site is large. On it, there are a lot of people, and working at heights is often seen, making it difficult to supervise and protect. So, it's necessary to install fences on the work site to limit the scope of operation personnel's movement. The fence shall be a closed mesh barrier with independent posts, and an appropriate number of sign boards reading "Stop! High voltage, danger!" should be hung around the fence, and the sign boards shall face what's inside the fence. This is to warn the maintainer to work only inside the fence. In order to make it easier for workers to access, the entrance of the fence should be arranged next to the adjacent road, and a sign board reading "Enter and exit from here" shall be provided.

To reduce the workload of the operator to provide safety measures while meeting the requirement of work safety, a fully enclosed fence shall be installed around the live equipment when the power to most outdoor equipment is cut off while other equipment is not likely to come in contact with the live conductors, and only a few locations remain live. At this time, the hazard is inside the fence. To prevent anyone from entering the electrified compartment by mistake, a sign board reading "Stop! High voltage, danger!" should be hung on the fence. Climbing the fence is prohibited. A sign board reading "Work here" shall be provided on the work site. To indicate an outdoor work site, a sign board reading "Work here" shall be provided at the equipment under maintenance to distinguish it from the operating equipment. If a work ticket specifies several work sites, each work site shall have a "Work here" sign board. When working at the AC/DC panel, protective panel, automation panel, etc., a "Work Here" sign board should be provided in front of and behind the panel.

When working on an outdoor framework, a sign board reading "Stop! High voltage, danger!" should be hung on the cross beam near the live part on the work site. A sign board saying "Get up and down from here" shall be hung on the staff-ascending iron stand or ladder. If other adjacent live frameworks are likely to be ascended by mistake, a sign board saying "Do not climb! High voltage, danger!" shall be hung. Because the cross beam of an outdoor framework is usually connected to an adjacent compartment, and it is difficult to distinguish adjacent live equipment when working at heights, it's likely that someone ascends a live equipment framework by mistake. Therefore, when working on an outdoor framework, a sign board reading "Stop! High voltage, danger!" should be hung on the cross beam near the live part on the work site, and the sign board shall face the working framework. Operators generally do not have any special climbing tools nor proficient climbing skills. So, when handling work permit procedures, the maintainer shall hang the sign boards under the supervision of the work permitter. And when handling the end-of-work procedures, they shall take down the sign boards under the supervision of the work permitter. To prevent workers from ascending a live framework, the operator shall hang a sign board saying "Get up and down from here" on the staff-ascending iron stand or ladder. To warn workers not to climb onto a live equipment framework, a sign board saying "Do not climb! High voltage, danger!" shall be hung on other nearby live frameworks that are likely to be ascended by mistake.

Workers are forbidden to move or remove any barriers (fences) and sign boards without permission. If it is necessary to move or remove barriers (fences) and sign boards for a short time as required, the permission of the work permitter shall be obtained, and the barriers (fences) and sign boards shall be moved or removed under the supervision of the person in charge of work and shall be re-installed immediately after the work is done. Temporary barriers, grounding wires, sign boards, and fences are measures taken to guarantee the safety of the maintainer and the safe operation of equipment. They are used to separate the work site from the electrified compartment to prevent workers from touching live equipment by mistake. So, changing safety measures without permission may change the scope of work and causes people to get electric shocks. If a worker needs to change the safety measures for a short time due to work needs, he/she shall obtain the permission of the work permitter and change the safety measures under the supervision of the person in charge of work. After completing the work, he/she shall immediately restore the safety measures and report to the work permitter. When someone needs to change or take new safety measures because of extending a work task, he/she must complete a new work ticket and re-handle work permit formalities. When working at a DC convertor station with monopole power interruption, a fence shall be provided between the equipment in the bipole common area and the power interruption area, and a sign board reading "Stop! High voltage, danger!" should be hung on the fence (the sign board shall face the power-interrupted equipment) and at the gate of the operating valve hall. Set up "Work here" sign boards at the maintenance valve hall and DC field equipment. The sign board hung is as shown in Fig. 4-9.

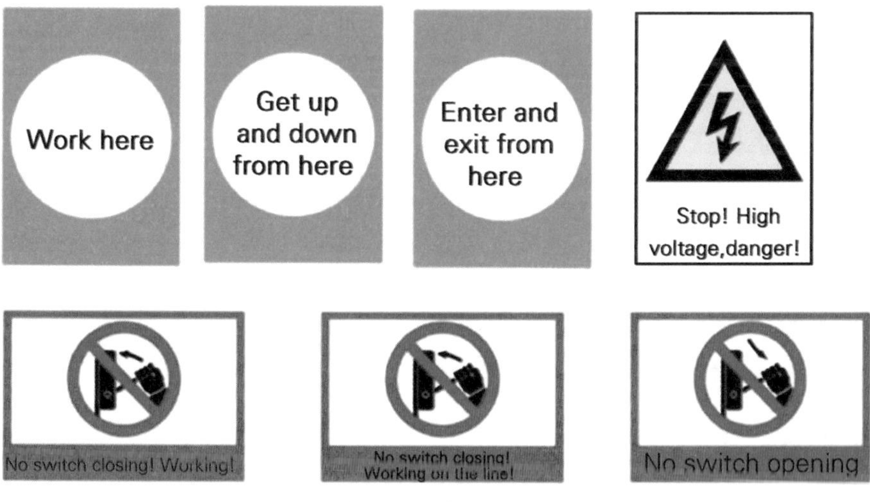

Fig. 4-9 Sign board

任务五　高压验电器的检查与使用实训

一、作业任务

（1）2人一组，分组完成任务。

学习正确检查10 kV（或35 kV）高压验电器外观。注意检查高压验电器有没有试验标签及出厂合格证。检查外观包括绝缘杆、护环、金属触头、试验按钮等，检查时每个部分都要检查，不能漏查或者不查。

（2）学习正确使用10 kV（或35 kV）高压验电器验电。

① 根据教师安排任务进行高压验电操作学习。首先，应根据被验电气设备的额定电压，选用合适型号的验电器，以免危及操作者人身安全或产生误判。

② 在指定作业场景验电，验电时应戴绝缘手套，手不超过握手的隔离护环。操作人员必须手握操作手柄，并将操作杆全部拉出定位后，方可按有关规定顺序进行验电操作（注意操作中是将验电器渐渐移向设备，在移近过程中若有发光或发声指示，则立即停止验电）。验电前应对设备逐相充分放电，先对离人体近的一相放电，再对离人体远的放电。

③ 在非全部停电场合进行验电操作，应先将验电器在有电部位上测试，再到施工部位进行测试，然后回复到有电部位上复测，以确保安全。不得以验电器的自检按钮试验替代本项操作。

④ 在全部停电场合进行验电操作时，应在验电操作前使用生产的相应电压等级的“验电信号发生器”对该验电器进行完好性的验证后，方可进行验电操作。

⑤ 若发现验电器欠压指示灯点亮时，则提示需立即更换新电池后再继续试验。

（3）任务结束后，收拾好工作现场，安全工器具归位。

二、引用标准及文件

（1）《国家电网公司电力安全工作规程（变电部分）》。

（2）《特种作业（电工）安全技术培训大纲和考核标准》。

三、作业条件

作业人员精神状态良好，熟悉工作中安全措施、技术措施以及现场工作危险点。

四、作业前准备

1. 现场作业基本要求及条件

勘察现场设备情况，查阅相关技术资料，包括历史数据及相关规程。

2. 工器具及材料选择

10 kV（或35 kV）高压验电器、绝缘手套、绝缘靴、停电设备、放电棒、安全帽、垫布。

3. 危险点及预防措施

（1）存放不当造成仪器仪表损坏。

危险点：不正确保存造成安全工器具损坏。

预防措施：10 kV（或 35 kV）高压验电器、绝缘手套、绝缘靴、放电棒、安全帽等使用前整齐安放在干净的垫布上，工器具不能叠放。

（2）作业人员伤害。

预防措施：正确按照安全措施要求进行操作。

3. 作业人员分工

现场工作负责人（监护人）：×××

现场作业人员：×××

五、作业规范及要求

（1）现场操作时，由老师现场指出验电的设备，学员按现场规程模拟演示。

（2）严格按照规程要求分步骤进行高压验电。

（3）必须按操作规程严格执行，一旦发生异常立刻停止作业并报告。

六、作业流程及标准（表 4-3）

表 4-3　高压验电器的检查、使用流程及考核评分标准

班级		姓名		学号		考评员		成绩	

序号	作业名称	质量标准	分值/分	扣分标准	扣分	得分
1	高压验电器的检查					
1.1	验电器选择	检查验电器：应使用相应电压等级且合格的接触式验电器	10	未检查或者选择错误电压等级验电器扣 10 分		
1.2	外观检查	检查高压验电器标签、合格证、外观	20	未检查标签扣 10 分，未检查合格证扣 5 分，未检查外观扣 5 分		
1.3	工作触头检查	通过按压工作触头，初步检查验电器合格	10	未通过按压，初步检查验电器，扣 10 分		
2	用高压验电器验电					
2.1	工作前准备	穿戴好工作服，戴绝缘手套，应将外衣袖口放入绝缘手套的伸长部分；穿绝缘靴，应将裤管套入靴筒内	10	未穿戴好工作服，未戴绝缘手套（未将外衣袖口放入绝缘手套的伸长部分），未穿绝缘靴（未将裤管套入靴筒内）每项扣 3 分		
2.2	放电	验电前用放电棒对设备逐相充分放电，先对离人体近的一相放电，再对离人体远的放电	10	未执行或执行不规范扣 10 分		

续表

序号	作业名称	质量标准	分值/分	扣分标准	扣分	得分
2.3	验电	证实验电器良好； 验电器的工作触头将在带电的设备上验电，不能直接接触带电体，只能逐渐接近带电体，至验电器发出声、光或其他报警信号为止； 验电器的伸缩式绝缘棒长度应拉足，验电时应手拿绝缘棒的握手部分，手不可超出护环，人体应与验电设备保持安全距离； 在指定停电设备上逐相进行验电（不能只验一相，以防某一相仍然有电压，发生触电事故），验电时，先验离人体近的一相，由近及远； 最后回到有电设备上复测，以确保安全； 任务结束后，收拾好工作现场，安全工器具归位	40	未将验电器的工作触头在带电的设备上验电，直接接触带电体，扣5分； 未逐渐接近带电体，至验电器发出声、光或其他报警信号为止，扣5分； 验电器的伸缩式绝缘棒长度未拉足，扣5分； 验电时应手未拿绝缘棒的握手部分，手超出护环，人体应与验电设备未保持安全距离，扣5分； 未在指定停电设备上逐相进行验电，只验一相，扣5分； 验电时，未先验离人体近的一相，由近及远，扣5分；最后未回到有电设备上复测，扣5分； 任务结束后，未收拾好工作现场，未将安全工器具归位，扣5分		
合计			100			

Task Ⅴ Inspection and training in using HV electroscope

Ⅰ. Operating tasks

(1) Two persons are grouped together to perform the task.

Learn to correctly check the appearance of a 10 kV (or 35 kV) HV electroscope. Be sure to check if the HV electroscope comes with a testing label and a certificate of conformity. Appearance inspection includes the insulating rod, guard ring, metallic contact terminal, test button, etc. Each part shall be checked. It's not allowed to omit, either deliberately or accidentally, any check items.

(2) Learn to correctly use a 10 kV (or 35 kV) HV electroscope.

① Learn how to perform HV electricity detection according to the task assigned by the instructor. First, select an appropriate model of electroscope according to the rated voltage of the electrical equipment to be detected, so as not to endanger the operator or make a misjudgment.

② When detecting electricity in a specified scenario, wear insulating gloves and do not reach over the guard ring of the handle. The operator must hold the operating handle, pull out the operating rod completely and fix it before detecting electricity in the prescribed order (be sure to move the electroscope gradually toward the equipment to detect electricity, and if there are any visual or audible indications in the approaching process, stop detecting immediately). Prior to the electricity detection, the equipment shall be fully discharged phase by phase, with the phase closer to the human body prioritized over a distant phase.

③ When it comes to detecting electricity in a scenario where not all parts are power-interrupted, an electroscope shall be used to detect the live part, the construction part, and then the live part again to ensure safety. This step shall not be replaced by the built-in test button testing of the electroscope.

④ When it comes to detecting electricity in a scenario where all parts are power-interrupted, a manufactured "electricity detection signal generator" of the corresponding voltage class shall be used to verify the intactness of the electroscope before electricity detection.

⑤ If the electroscope's undervoltage indicator light is on, the battery should be replaced by a new one immediately before continuing the test.

(3) After the task, clean the work site and return the safety tools and instruments to their original positions.

Ⅱ. Referenced standards and documents

(1) *Electric Power Safety Working Regulations (Power Transformation) of State Grid Corporation of China*;

(2) *Safety Technology Training Outline and Appraisal Standards for Special Operation (Electrician)*.

Ⅲ. Operating conditions

The operation personnel shall be in a good mental state and familiar with safety measures, technical measures, and dangerous points of field work.

Ⅳ. Preparation before operation

1. Basic requirements and conditions for on-site operations

Conduct a survey of equipment on site and refer to relevant technical data, including historical data and related procedures.

2. Selection of tools and instruments and materials

10 kV (or 35 kV) HV electroscope, insulating gloves, insulating boots, power-interrupted equipment, discharging rod, safety helmet, placemat.

3. Dangerous points and preventive measures

(1) Instrument damage due to improper storage.

Dangerous points: Safety tools and instruments will be damaged if not stored properly.

Prevention and control measures: The 10 kV (or 35 kV) HV electroscope, insulating gloves, insulating boots, discharging rods, safety helmets, etc. shall be neatly put on a clean placemat before use. The tools and instruments shall not be stacked.

(2) Injury to the operator.

Prevention and control measures: Operate correctly according to the requirements of safety measures.

4. Division of labor among operators

Person in charge of on-site work (supervisor): ×××

On-site operator: ×××

Ⅴ. Operating specifications and requirements

(1) When it comes to on-site operation, the instructor shall specify the equipment to be detected, and the trainees shall simulate the demonstration according to the on-site procedures.

(2) HV electricity detection shall be performed step by step in strict accordance with the procedures.

(3) The operating procedures must be strictly followed Stop immediately and report in the event of any anomaly.

Ⅵ. Operation process and standards (see Tab. 4-3)

Tab. 4-3 Inspection and use process of HV electroscope, and assessment and scoring standards

Class		Name		Student ID		Examiner		Score	

S/N	Operation name	Quality standard	Points	Deduction criteria	Deduction	Score
1	Inspection of HV electroscope					
1.1	Electroscope inspection	Check the electroscope: An compliant contact electroscope of an appropriate voltage class should be used	10	Deduct 10 points if failing to check or select an electroscope of an improper voltage class		
1.2	Visual inspection	Check the label, certificate of conformity and appearance of the HV electroscope	20	Deduct 10 points if failing to check the label and 5 points if failing to check the certificate of conformity or the appearance		
1.3	Contact terminal inspection	Initially check if the electroscope is compliant by pressing the working contact terminal	10	Deduct 10 points if failing to initially check the electroscope by pressing		
2	Electricity detection by HV electroscope					
2.1	Preparations before work	Wear good work clothes and insulating gloves: the coat cuffs should be tucked into the extended part of the insulating gloves. Put on insulating boots, and the trouser legs shall be tucked into the boots	10	Deduct 3 points if failing to properly wear or wear work clothes, insulating gloves (the coat cuffs should be tucked into the extended part of the insulating gloves), or insulating boots (the trouser legs shall be tucked into the boots)		
2.2	Discharging	Prior to the electricity detection, the equipment shall be fully discharged phase by phase using a discharging rod, with the phase closer to the human body prioritized over a distant phase	10	Deduct 10 points if failing to do so or failing to do it properly		
2.3	Electricity detection	Verify if the electroscope is in good condition. The working contact terminal of the HV electroscope shall detect the live equipment. The electroscope shall not directly contact the charged body, but shall only be gradually moved towards the charged body until the electroscope emits sound, light, or other alarm signals are given. The retractable insulating rod of the electroscope shall be extended long enough. When detecting electricity, you should hold the handle and not reach over the guard ring of the insulating rod, and you should maintain a safe distance from the equipment to be detected.	40	Deduct 5 points if the electroscope comes in direct contact with the charged body instead of detecting the live equipment with the working contact terminal of the electroscope; Deduct 5 points if failing to move the electroscope towards the charged body until the electroscope emits sound, light, or other alarm signals are given; Deduct 5 points if failing to extend the retractable insulating rod of the electroscope long enough;		

Continued

S/N	Operation name	Quality standard	Points	Deduction criteria	Deduction	Score
2.3	Electricity detection	Detect the designated power-interrupted equipment phase by phase (the operator shall not detect only one phase lest any electric shock should happen if another phase retains the voltage). Electricity shall be detected from near to far, with the phase closer to the human body detected first. At last, re-detect the live equipment to ensure safety. After the task, clean the work site and return the safety tools and instruments to their original positions		Deduct 5 points if failing to hold the handle of the insulating rod when detecting electricity, reach over the guard ring, or maintain a safe distance between the human body and the equipment to be detected; Deduct 5 points if failing to detect one phase, instead of detecting the power-interrupted equipment phase by phase; Deduct 5 points if failing to detect electricity from near to far, instead of detecting the phase closer to the human body first; Deduct 5 points if failing to re-detect the live equipment; Deduct 5 points if failing to clean the work site and return the safety tools and instruments to their original positions after the task		
Total			100			

任务六　装拆接地线实训

一、作业任务

（1）2 人一组，分组完成任务。

（2）学习正确检查 10 kV（或 35 kV）三相成套短路接地线外观。注意检查 10 kV（或 35 kV）三相成套短路接地线有没有试验标签及出厂合格证。检查外观包括绝缘杆、接地线（必须是多股软裸铜线，横截面积不得小于 25 mm^2）、金属触头等，检查时每个部分都要检查，不能漏查或不查。

（3）根据老师安排任务进行装拆接地线操作。

① 根据被验电气设备的额定电压，选用合适型号的三相成套短路接地线，以免危及操作者人身安全；每组接地线均应编号。

② 工作人员两人，一人监护，一人操作，操作人员穿戴好绝缘安全防护用具（安全帽、绝缘靴、绝缘手套）。

③ 在验明确无电压的设备上装设接地线。按先接接地端，再接设备端进行。接地线必须用专用线夹将其固定在导线上，严禁用缠绕方式连接，保证接地线接触良好。装设导体端时，先装设离人体近的一相，由近及远，必须接触牢靠。

④ 拆除接地线时按先拆设备端，再拆接地端进行。

（4）任务结束后，收拾好工作现场，安全工器具归位。

二、引用标准及文件

（1）《国家电网公司电力安全工作规程（变电部分）》。

（2）《特种作业（电工）安全技术培训大纲和考核标准》。

三、作业条件

作业人员精神状态良好，熟悉工作中安全措施、技术措施以及现场工作危险点。

四、作业前准备

1. 现场作业基本要求及条件

勘察现场设备情况，查阅相关技术资料，包括历史数据及相关规程。

2. 工器具及材料选择

10 kV（或 35 kV）三相成套短路接地线、绝缘手套、绝缘靴、已验明停电设备、安全帽、垫布。

3. 危险点及预防措施

（1）存放不当造成安全工器具损坏。

危险点：不正确保存造成安全工器具损坏。

预防措施：10 kV（或 35 kV）三相成套短路接地线、绝缘手套、绝缘靴、安全帽等使用前整齐安放在干净的垫布上，工器具不能叠放。

（2）作业人员伤害。

预防措施：正确按照安全措施要求进行操作。

4. 作业人员分工

现场工作负责人（监护人）：×××

现场作业人员：×××

五、作业规范及要求

（1）现场操作时，由老师现场指定已验明无电设备，学员按现场规程模拟演示。

（2）严格按照规程要求分步骤进行装挂接地线。

（3）必须按操作规程严格执行，一旦发生异常立刻停止作业并报告。

六、作业流程及标准（表 4-4）

表 4-4　装拆接地线流程及考核评分标准

班级		姓名	学号	考评员		成绩	
序号	作业名称	质量标准		分值/分	扣分标准	扣分	得分
1	接地线的选择及外观检查	选择型号合适的三相成套短路接地线； 检查三相成套短路接地线有没有试验标签及出厂合格证，检查外观包括绝缘杆、接地线（必须是多股软裸铜线，截面面积不得小于 25 mm²）、金属触头		20	未选择型号合适的三相成套短路接地线，扣 10 分； 未检查三相成套短路接地线有没有试验标签及出厂合格证，未检查外观包括绝缘杆、接地线（必须是多股软裸铜线，截面面积不得小于 25 mm²）、金属触头等扣 10 分		
2	装接地线	每组接地线均应编号，工作人员两人，一人监护，一人操作，操作人员穿戴好绝缘安全防护用具（安全帽、绝缘靴、绝缘手套），在验明确无电压的设备上装设接地线；按先接接地端、再接设备端进行，接地线必须用专用线夹将其固定在导线上，严禁用缠绕方式连接；保证接地线接触良好，装设导体端时，先装设离人体近的一相，由近及远，必须接触牢靠		60	每组接地线未编号，扣 10 分； 操作人员未穿戴好绝缘安全防护用具（安全帽、绝缘靴、绝缘手套），扣 10 分；未在验明确无电压的设备上按先接接地端、再接设备端进行，扣 15 分； 接地线未用专用线夹而是用缠绕方式连接，扣 15 分； 装设导体端时，未先装设离人体近的一相，由近及远，扣 10 分		

续表

序号	作业名称	质量标准	分值/分	扣分标准	扣分	得分
3	拆接地线	拆除接地线时按先拆设备端、再拆接地端进行； 任务结束后，收拾好工作现场，安全工器具归位	20	拆除接地线时先拆接地端、再拆设备端，扣15分； 任务结束后，未收拾好工作现场，安全工器具未归位，扣5分		
合计			100			

Task Ⅵ Training in mounting and removing grounding wires

Ⅰ. Operating tasks

(1) Two persons are grouped together to perform the task.

(2) Learn to correctly check the appearance of a 10 kV (or 35 kV) three-phase short-circuit grounding wire set. Be sure to check if a 10 kV (or 35 kV) three-phase short-circuit grounding wire set comes with a testing label and a certificate of conformity. Appearance inspection shall include the insulating rods, grounding wires (which must be multistranded, bare, and annealed copper wires with a cross sectional area no less than 25 mm^2), and metallic contact terminals. Each part shall be checked. It's not allowed to omit, either deliberately or accidentally, any check items.

(3) Mount and remove grounding wires according to the task assigned by the instructor.

① Select an appropriate model of three-phase short-circuit grounding wire set according to the rated voltage of the electrical equipment to be detected, so as not to endanger the operator. Each group of grounding wires shall be numbered.

② Two workers cooperate with each other, with one supervising and another operating. The operator shall wear the insulating safety protective appliances (safety helmet, insulating boots, and insulating gloves).

③ Mount the grounding wire on the provenly voltage-free equipment. Connect the grounding terminal first and then the equipment terminal. The grounding wire must be fixed on the conductor with a special wire clamp, and it is strictly prohibited to connect by winding to ensure that the grounding wire is in good contact. When connecting to the conductor terminal, it must be installed from near to far, with the phase closer to the human body to be installed first, and it must be in firm contact.

④ To remove the grounding wire, remove the equipment terminal and then the grounding terminal.

(4) After the task, clean the work site and return the safety tools and instruments to their original positions.

Ⅱ. Referenced standards and documents

(1) *Electric Power Safety Working Regulations (Power Transformation) of State Grid Corporation of China*;

(2) *Safety Technology Training Outline and Appraisal Standards for Special Operation (Electrician)*.

Ⅲ. Operating conditions

The operation personnel shall be in a good mental state and familiar with safety measures, technical measures, and dangerous points of field work.

Ⅳ. Preparation before operation

1. Basic requirements and conditions for on-site operations

Conduct a survey of equipment on site and refer to relevant technical data, including historical data and related procedures.

2. Selection of tools and instruments and materials

10 kV (or 35 kV) three-phase short-circuit grounding wire set, insulating gloves, insulating boots, provenly power-interrupted equipment, safety helmet, and placemat.

3. Dangerous points and preventive measures

(1) Safety tools and instruments damaged by improper storage.

Dangerous points: Safety tools and instruments will be damaged if not stored properly.

Prevention and control measures: The 10 kV (or 35 kV) three-phase short-circuit grounding wire set, insulating gloves, insulating boots, safety helmets, etc. shall be neatly put on a clean placemat before use. The tools and instruments shall not be stacked.

(2) Injury to the operator.

Prevention and control measures: Operate correctly according to the requirements of safety measures.

4. Division of labor among operators

Person in charge of on-site work (supervisor): ×××

On-site operator: ×××

Ⅴ. Operating specifications and requirements

(1) When it comes to on-site operation, the instructor shall point out the provenly power-interrupted equipment, and the trainees shall simulate the demonstration according to the on-site procedures.

(2) The grounding wires shall be hung step by step in strict accordance with the procedures.

(3) The operating procedures must be strictly followed Stop immediately and report in the event of any anomaly.

Ⅵ. Operation process and standards (see Tab. 4-4)

Tab. 4-4 Mounting and removal process of grounding wires, and assessment and scoring standards

Class		Name		Student ID		Examiner		Score	
S/N	Operation name	Quality standard			Points	Deduction criteria		Deduction	Score
1	Selection and appearance inspection of grounding wires	Select a three-phase short-circuit grounding wire set of a suitable model; Check if the three-phase short-circuit grounding wire set comes with a testing label and a certificate of conformity. Appearance inspection shall include the insulating rods, grounding wires (which must be multistranded, bare, and annealed copper wires with a cross sectional area no less than 25 mm^2), and metallic contact terminals			20	Deduct 10 points if failing to elect a three-phase short-circuit grounding wire set of a suitable model; Deduct 10 points if failing to check if the three-phase short-circuit grounding wire set comes with a testing label and a certificate of conformity. Appearance inspection shall include the insulating rods, grounding wires (which must be multistranded, bare, and annealed copper wires with a cross sectional area no less than 25 mm^2), and metallic contact terminals			
2	Grounding wire mounting	Each group of grounding wires should be numbered. Two workers cooperate with each other, with one supervising and another operating. The operator shall wear the insulating safety protective appliances (safety helmet, insulating boots, and insulating gloves) and mount grounding wires on the provenly voltage-free equipment. Connect the grounding terminal and the equipment terminal. The grounding wires must be fixed on the conductor with a special wire clamp, and it is forbidden to connect by winding. Make sure that the grounding wire is in good contact. When connecting to the conductor terminal, it must be installed from near to far, with the phase closer to the human body to be installed first, and it must be in firm contact. To remove the grounding wire, remove the equipment terminal and then the grounding terminal			60	Deduct 10 points if failing to number each group of grounding wires; Deduct 10 points if the operator fails to wear the insulating safety protective appliances (safety helmet, insulating boots, and insulating gloves); Deduct 15 points if failing to connect the grounding terminal before the equipment terminal on the provenly voltage- free equipment; Deduct 15 points if connecting the grounding wire by binding, instead of fixing it with a special wire clamp; Deduct 10 points if failing to mount the conductor terminal from near to far with the phase closer to the human body to be connected first			

Continued

S/N	Operation name	Quality standard	Points	Deduction criteria	Deduction	Score
3	Grounding wire removal	After the task, clean the work site and return the safety tools and instruments to their original positions	20	Deduct 15 points if removing the grounding terminal before the equipment terminal when removing the grounding wire; Deduct 5 points if failing to clean the work site and return the safety tools and instruments to their original positions after the task		
Total			100			

任务七　悬挂标示牌和装设遮栏实训

一、作业任务

（1）2 人一组，分组完成任务。

根据老师安排任务进行悬挂标示牌和装设遮栏操作。

（2）学习常用安全标示牌辨识，教师任意选择 5 个安全标示牌，由学员指认。学员能对指定的安全标示牌用途进行说明，并解释其用途。

按照指定的作业场景，装设遮栏，并正确布置相关的安全标示牌 3 个。

（3）任务结束后，收拾好工作现场，安全工器具归位。

二、引用标准及文件

《国家电网公司电力安全工作规程（变电部分）》。

《特种作业（电工）安全技术培训大纲和考核标准》。

三、作业条件

作业人员精神状态良好，熟悉工作中安全措施、技术措施以及现场工作危险点。

四、作业前准备

1. 现场作业基本要求及条件

勘察现场设备情况，查阅相关技术资料，包括历史数据及相关规程。

2. 工器具及材料选择

常用安全标示牌 5 种，遮栏 4 块。

3. 危险点及预防措施

（1）存放不当造成标示牌损坏。

危险点：不正确存放造成遮栏及标示牌损坏。

预防措施：遮栏及所有标示牌分类整齐存放。

（2）作业人员伤害。

预防措施：正确按照安全措施要求进行操作。

4. 作业人员分工

现场工作负责人（监护人）：×××

现场作业人员：×××

五、作业规范及要求

（1）现场操作时，由老师现场指定已装挂接地线的设备，学员按现场规程模拟演示。

（2）严格按照规程要求进行装设遮栏和悬挂标示牌。
（3）必须按操作规程严格执行，一旦发生异常立刻停止作业并报告。

六、作业流程及标准（表 4-5）

表 4-5　悬挂标示牌和装设遮栏流程及考核标准

班级		姓名		学号		考评员		成绩	

序号	作业名称	质量标准	分值/分	扣分标准	扣分	得分
1	装设遮栏	按指定的作业场景装设遮栏；遮栏摆放位置合理，留有进出口通道	10	遮栏摆放位置不合理，未留有进出口通道，扣 10 分		
2	悬挂标示牌					
2.1	熟悉常用安全标识牌	指认教师指定的安全标示（5 个）	20	错误指认教师指定的安全标示（5 个），每错一个扣 4 分		
2.2	常用安全标示用途解释	能对指定的安全标示（5 个）进行用途说明，并解释其作用	20	未能对指定的安全标示（5 个）进行用途说明，并解释其作用，错一个扣 4 分		
2.3	正确悬挂布置安全标示	按指定的作业场景，正确布置安全标示（3 个）	50	未按指定的作业场景正确布置安全标示（3 个），选错标识 1 个扣 10 分，摆放位置错误 1 个扣 10 分		
合计			100			

Task Ⅶ Training in hanging sign boards and mounting barriers

Ⅰ. Operating tasks

(1) Two persons are grouped together to perform the task.

Hang sign boards and mount barriers according to the task assigned by the instructor.

(2) Learn to distinguish common safety sign boards. The instructor chooses 5 safety sign boards randomly and the trainees identify them.

The trainees shall be able to explain the designated purpose of the safety sign boards.

The trainees shall mount barriers and set up 3 appropriate safety sign boards properly according to the specified work scenarios.

(3) After the task, clean the work site and return the safety tools and instruments to their original positions.

Ⅱ. Referenced standards and documents

(1) *Electric Power Safety Working Regulations (Power Transformation) of State Grid Corporation of China*;

(2) *Safety Technology Training Outline and Appraisal Standards for Special Operation (Electrician)*.

Ⅲ. Operating conditions

The operation personnel shall be in a good mental state and familiar with safety measures, technical measures, and dangerous points of field work.

Ⅳ. Preparation before operation

1. Basic requirements and conditions for on-site operations

Conduct a survey of equipment on site and refer to relevant technical data, including historical data and related procedures.

2. Selection of tools and instruments and materials

5 types of common safety sign boards, and 4 barriers.

3. Dangerous points and preventive measures

(1) Sign board damage due to improper storage.

Dangerous points: Barriers and sign boards will be damaged if not stored properly.

Prevention and control measures: All barriers and sign boards shall be neatly stored by classifications.

(2) Injury to the operator.

Prevention and control measures: The operator shall abide by the requirements of safety measures.

4. Division of labor among operators

Person in charge of on-site work (supervisor): ×××

On-site operator: ×××

Ⅴ. Operating specifications and requirements

(1) When it comes to on-site operation, the instructor shall point out the equipment that a grounding wire is hung on, and the trainees shall simulate the demonstration according to the on-site procedures.

(2) The barriers and sign boards shall be set up in strict accordance with the procedures.

(3) The operating procedures must be strictly followed Stop immediately and report in the event of any anomaly.

Ⅵ. Operation process and standards (see Tab. 4-5)

Tab. 4-5 Process for hanging sign boards and mounting barriers and appraisal standards

Class		Name		Student ID		Examiner		Score	
S/N	Operation name	Quality standard		Points	Deduction criteria			Deduction	Score
1	Barrier mounting	Mount barriers according to the designated working scenario. The barriers shall be reasonably arranged with an access way.		10	Deduct 10 points if the barriers are not reasonably arranged and have no access way				
2	Sign board hanging								
2.1	Being familiar with common safety signs	Identify the 5 safety signs specified by the instructor		20	Deduct 4 points if failing to identify any of the 5 safety signs specified by the instructor				
2.2	Explaining of purposes of common safety signs	Be able to explain the purposes of the 5 designated safety signs		20	Deduct 4 points if failing to explain the purposes of any of the 5 designated safety signs				
2.3	Correctly hanging safety signs	Correctly hanging 3 safety signs the according to the designated working scenarios		50	In terms of failing to correctly hang 3 safety signs the according to the designated working scenarios, deduct 10 points if choosing a wrong safety sign and 10 points if hanging the safety sign at a wrong position				
Total				100					

参考文献

[1] 顾飚．电力安全知识及案例[M]．北京：中国电力出版社，2015．
[2] 许庆海．电力安全基本技能[M]．北京：中国电力出版社，2008．
[3] 张良瑜．电业安全[M]．北京：中国电力出版社，2012．
[4] 中国安全生产协会注册安全工程工作委员会．安全生产技术 2008 年版[M]．中国大百科全书出版社，2008．
[5] 国家电网公司．国家电网公司电力安全工作规程（变电部分）[M]．北京：中国电力出版社，2009．